# TRAVELLERS
## —IN THE—
# NEAR EAST

*With warm thanks for your help,*

*Brenda Moon*

# TRAVELLERS
— IN THE —
NEAR EAST

*Editor*
Charles Foster

STACEY INTERNATIONAL

Travellers in the Near East

Copyright © Stacey International & ASTENE

First printed 2004
Reprinted 2006

The Association for the Study of Travellers in Egypt and the Near East
*with*
Stacey International
128 Kensington Church Street
London W8 4BH
Tel: 020 7221 7166 Fax: 020 7792 9288

www.stacey-international.co.uk

ISBN: 1900988 712

CIP Data: A catalogue record for this book is available from the British Library

*Design:* Kitty Carruthers

*Additional editing:* Caroline Singer

*Printing & Binding:* Lightning Source UK

All rights reserved. No part of this publication may be reproduced, stored in a retrieval system, or transmitted in any form or by any means, electronic, mechanical, photocopying, recording, or otherwise, without the prior permission of the copyright owners.

# Contents

**List of Illustrations**   7
**Contributors' Biographies**   11

**Publisher's Note**   15

**Chapter One**   17

*A Tale of Two Ciceros: Travels in Asia Minor
   in the late Roman Republic*
**Marsha B. McCoy,** Department of Classical Studies,
Fairfield University, Connecticut.

**Chapter Two**   29

*Mercantile Gentlemen and Inquisitive Travellers:
   Constructing* The Natural History of Aleppo
**Dr Janet Starkey,** University of Durham.

**Chapter Three**   71

*Jean-Baptiste Adanson (1732-1804): A French Dragoman in
   Egypt and the Near East*
**Jen Kimpton,** Johns Hopkins University.

**Chapter Four**   105

*The Journey of the Comte de Forbin in the Near East and
   Egypt, 1817-1818*
**Pascale Linant de Bellefonds,** Paris.

**Chapter Five**   133

*Travellers, Tribesmen and Troubles: Journeys to Petra, 1812-1914*
**Norman N. Lewis,** formerly principal instructor at the Middle
East Centre for Arab Studies, Lebanon.

**Chapter Six** 153

*Surveying the Morea: The French Expedition, 1828-1832*
**Professor Malcolm Wagstaff**, Southampton University.

**Chapter Seven** 165

*La Mission Scientifique de Morée: Captain Peytier's Contribution*
**Dr Elizabeth French**, formerly Director of British School of Athens, and **William M. Frick**, Connecticut.

**Chapter Eight** 181

*Christian Rassam (1808-1872): Translator, Interpreter, Diplomat and Liar*
**Dr Geoffrey Roper**, University Library, Cambridge.

**Chapter Nine** 199

*Mr and Mrs Smith in Greece, Egypt and the Levant*
**Dr Brenda E. Moon**, formerly University Librarian, Edinburgh University.

**Chapter Ten** 223

*Robert Murdoch Smith and the Mausoleum: Excavations at Halicarnassus (Bodrum) 1856-1859*
**Jennifer Scarce**, University of Dundee.

**Chapter Eleven** 245

*Listening for 'the Sound of Running History': Sir George Adam Smith, 1856-1942*
**Rev Iain D. Campbell**, Ph.D., Edinburgh.

**Chapter Twelve** 287

*Politics and the Travels of Gertrude Bell*
**Dr Richard Long**, University of Newcastle upon Tyne.

# List of Illustrations

Page

19 **Figure 1:** *Cistophorus* ('cistophoric' tetradrachm, the normal silver currency of Asia Minor, so called because it usually had a representation of the sacred *cists* or chest of Dionysus between guardian snakes, also attributes of Dionysus, who was traditionally from Phrygia). It was struck at Apamea in Phrygia, part of the province of Cilicia 56-50 BC, during Marcus Cicero's governorship. It carries the legend on the top of the coin in Latin: CICERO M.F. PROCOS (Cicero, son of Marcus, proconsul). The legend on the bottom of the coin is in Greek.

24 **Figure 2:** Map of Asia Minor in the time of the Ciceros.

44 **Figure 3:** *A Plan of the City of Aleppo* drawn by Carsten Niebuhr, printed opposite page 13 in *The Natural History of Aleppo* (1794), Vol. I.

45 **Figure 4:** Aleppine merchants and officials. Plate II in *Aleppo* (1794), I, opposite p 102.

86 **Figure 5:** Fragment of an Egyptian statue from *Origine des Egyptiens et de leur écriture symbolique*, Jean-Baptiste Adanson (Plate 17).

87 **Figure 6:** Hieroglyphic inscriptions in *Origine des Egyptiens et de leur écriture symbolique*, Jean-Baptiste Adanson (Plate 24).

116 **Figure 7:** The Comte de Forbin, by Paulin Guérin.

117 **Figure 8:** Ephesus, the Persecution Gate (after *Voyage dans le Levant*, pl. 6).

117 **Figure 9:** General view of Jerusalem (after *Voyage dans le Levant*, pl. 17).

118 **Figure 10:** Cairo, the caliphs' tombs (after *Voyage dans le Levant*, pl. 49).

118  **Figure 11:** Karnak, the great temple of Amon (after *Voyage dans le Levant,* pl. 60).

119  **Figure 12:** Drovetti, Forbin and the French team in Thebes, by Granger (after *Voyage dans le Levant,* pl. 73).

119  **Figure 13:** Alexandria, Cleopatra's Needles (after *Voyage dans le Levant,* pl. 76).

120  **Figure 14:** Muhammad Ali, drawing by Forbin (after G. Wiet, *Mohammed Ali et les Beaux-Arts* [Le Caire] pl. XLII).

121  **Figure 15:** The Mamelukes' massacre, by Horace Vernet (after *Voyage dans le Levant,* pl. 55).

138  **Figure 16:** A typical group of Petra Bedouin c. 1895: from *Petra: Perea Phoenicia*, Rev. A. Forder, Marshall Brothers, London and Edinburgh. Book undated, but probably published about 1920, p. 15.

138  **Figure 17:** Petra guides: from *Petra: Perea Phoenicia*, Rev A. Forder, p. 15.

139  **Figure 18:** Title page of Formby's *A Visit to the East*, 1843, with a sketch of Sheikh Muqabil abou Zaitoun 'Chief of the Inhabitants of Wadi Mousa'. Formby was at Petra in the spring of 1840; Muqabil died two or three years later.

140  **Figure 19:** Photograph of the Eastern Cliff at Petra taken by John Shaw Smith in 1852. Reproduced with the permission of the Palestine Exploration Fund.

142  **Figure 20:** The group of artists and others led by Jean-Leon Gerome on his 'Expedition en Orient' in 1868. The picture was painted by W. de Famars Testas. Reproduced by permission of the Rijksmuseum van Oudheden, Leiden.

144  **Figure 21:** Map showing the location of Petra.

158  **Figure 22:** An extract from the *Carte de la Morée* by Jean-Denis Barbié du Bocage, drawn and engraved in 1807 but not published until 1814.

159  **Figure 23:** An extract from the *Carte de la Morée* produced by the Geographical Engineers attached to the French Expedition to the Morea and published in 1832.

174  **Figure 24:** 'Drilling the Typikon'.

175  **Figure 25:** 'Palamide, as from our lodgings (at Nauplion)'.

186  **Figure 26:** *Kitāb Siyāḥat al-Masīḥī*: Malta 1834. Bunyan's *Pilgrim's Progress*, translated into Arabic by Christoph Schlienz and Christian Rassam. The wood-cut vignette depicts the author languishing in Bedford jail.

204  **Figure 27:** The Temple of Jupiter, Athens, October 1851.

205  **Figure 28:** A street in Constantinople (Istanbul), November 1851.

206  **Figure 29:** A house in Cairo, November 1851.

207  **Figure 30:** Kom Ombo, January 1852.

208  **Figure 31:** Temple of Isis, Philae, December 1851.

209  **Figure 32:** The Great Temple, Abu Simbel, December 1851.

209  **Figure 33:** From the walls of the Great Hall, Karnak, January 1852.

210  **Figure 34:** The Well of St Stephen and travellers' apartments, Sinai, March 1852.

211  **Figure 35:** Petra, March 1852.

212 **Figure 36:** Baalbek, May 1852.

230 **Figure 37:** Lieutenant Robert Murdoch Smith R.E. in 1856 (from a colour photograph by Alexander Stanesby).

231 **Figure 38:** Bay of Bodrum with the Castle of St Peter (author's photograph).

232 **Figure 39:** Marble statue of Mausolus c. 350 BC (British Museum, Greek and Roman Department, 1000, author's photograph).

233 **Figure 40:** Marble statue of Artemesia c. 350 BC (British Museum, Greek and Roman Department, 1001, author's photograph).

233 **Figure 41:** Marble lion from the Mausoleum c. 350 BC (British Museum, Greek and Roman Department, 1075, author's photograph).

234 **Figure 42:** Sir Robert Murdoch Smith in 1898 (from a photograph by R.S. Webster, Edinburgh).

Editor CHARLES FOSTER is a barrister, writer, traveller, and a Tutor in Medical Law and Ethics at the University of Oxford. He read law and veterinary medicine at Cambridge, and then worked on the immobilization of gazelles in Saudi Arabia, and the comparative anatomy of the Himalayan Hispid Hare at the Royal College of Surgeons. He was a Research Fellow at the Hebrew University, Jerusalem. He has numerous publications which have recently included pieces on wolf hunting in Kazakhstan, swimming the Hellespont, a number of essays on the literature of travel, and a chapter in the ASTENE book *Desert Travellers: from Herodotus to T.E. Lawrence* on 'The Zoology of Herodotus and his Greek descendants'. A lot of his life is spent on expeditions, particularly to desert regions such as the Western Desert, the Algerian Sahara, Sinai and the Danakil Depression. In 2002 he skied to the North Pole.

# *Contributors' Biographies*

REV DR IAIN D. CAMPBELL is a minister of the Free Church of Scotland, currently working in Back in the Isle of Lewis. He holds degrees from the Universities of Glasgow, London and Edinburgh, and received his ministerial training in the Free Church College, Edinburgh. The life and work of George Adam Smith was the subject of his doctoral studies. He is married to Anne, a teacher, and they have three children. He has written extensively on theological subjects, having published *The Doctrine of Sin* in 1999, as well as scholarly articles in the *Westminster Theological Journal* and the *Scottish Bulletin of Evangelical Theology*. He preaches and writes in both English and Gaelic.

DR ELIZABETH FRENCH is a former Director of the British School of Athens and a former Treasurer of ASTENE. She is an archaeologist and the daughter and wife of archaeologists. She has lived in Alexandria and worked for many years in Turkey and Greece. She has numerous publications to her name.

WILLIAM M. FRICK was born in Pennsylvania, and fought with the US Army in Europe during the Second World War. After the war he was involved in various business ventures and then began to study and teach Greek mythology. This led to many expeditions to Greece and a lasting fascination with Greek archaeology. He has written on a number of subjects, including Crusader castles in Argolis and 19th century travellers in the Plain of Argos. He now lives in Riverside, Connecticut.

JEN KIMPTON is a Ph.D. student of Egyptology in the Department of Near Eastern Studies at Johns Hopkins University, and an epigrapher/librarian for the Epigraphic Survey of the Oriental Institute, University of Chicago in Luxor, Egypt. Her dissertation is based on the works of Jean-Baptiste Adanson.

NORMAN N. LEWIS has been associated with the Near East for much of his life, first during the Second World War, then as principal instructor at the Middle East Centre for Arab Studies in Lebanon and later as an executive of a major oil company. He is the author of *Nomads and Settlers in Syria and Jordan, 1800-1980* (Cambridge, 1987) and of numerous articles and papers on different aspects of Syrian history.

PASCALE LINANT DE BELLEFONDS is a research fellow at the CNRS (Centre National de la Recherche Scientifique) and the Director of a research team of the CNRS in Nanterre University, near Paris. Her main interests are classical archaeology (particularly Greek and Roman iconography of mythology and religion), and the discovery of archaeological sites by 19th century travellers. She has travelled widely in the Near East, Egypt and Turkey.

DR C.W. RICHARD LONG holds degrees in Middle Eastern subjects from St Catharine's College, Cambridge, and McGill University. He worked for the Foreign Office (1962-9; British Embassy, Baghdad, 1963-6), the British Council (1969-90, in Beirut, Khartum, Ankara, Abu Dhabi, Amman, Doha and Teheran) and Newcastle Univeristy as Director of the

International Office (1990-5), later Associate Director (Middle East) and Director, Islamic Studies (1995-9). He taught at CMEIS, Durham University, 1965/6. His publications include *Tawfiq al Hakim, Playwright of Egypt* (Ithaca Press, 1979), *Bygone Heat: Travels of an Idealist in the Middle East* (Radcliffe Press, 2001, I.B. Taurus, 2004), *British Pro-Consuls in Egypt, 1914-1929, The Challenge of Nationalism* (the first volume of a trilogy on the British in the Middle East), journal articles, reviews and translations from Arabic and the text of the exhibition celebrating the 25th anniversary of the independence of the UAE.

MARSHA B. MCCOY holds degrees from Bryn Mawr College in Greek and in Classical and Near Eastern Archaeology; from Wadham College, Oxford University, in Literae Humaniores; from Harvard University in History; and from Yale University in History. She has held a Fulbright Fellowship at the University of Munich, Germany, a Summer Graduate Fellowship at the American Numismatic Society in New York City, and a Mellon Fellowship at New York University. She has taught at Harvard, Yale, and New York Universities, and currently teaches at Fairfield University in Connecticut. Her primary area of publication, and the subject of her PhD dissertation, is the political programme of the brothers Cicero in the late Roman Republic, but she also undertakes research in numismatics (the coinage of the Roman colony of Narbo Martius), Roman law (evidentiary presumptions in Roman law), Augustan culture (Virgil's *Aeneid* and the Temple of Mars Ultor in Augustan Rome), Petronius (Bakhtin and the *Satyria*), and Apuleius (scape-goating in the *Metamorphoses*).

DR BRENDA E. MOON was Librarian at the University of Edinburgh for seventeen years until her retirement in December 1996. She is a Fellow of the Royal Society of Edinburgh.

DR GEOFFREY ROPER was an Islamic bibliographer at the Cambridge University Library. He has lived in and travelled extensively through the Near and Middle East. He is the author or editor of numerous publications including *Index Islamicus* (of which he has been the editor since 1982), the *World Survey of Islamic Books* (editor 1992-1994), *European-language Periodicals as Sources of Information on the Muslim World* (1999) and *Middle Eastern Languages and the Print Revolution* (2002).

JENNIFER SCARCE was Curator of Middle Eastern Cultures at the National Museums of Scotland, and is now an Honorary Lecturer at the School of Design, Duncan of Jordanstone College of Art, University of Dundee, and works also as a research and travel consultant, freelance curator and author. Her many publications include a survey of Romanian carpet weaving, and articles and books on Kuwait, Iran, and Middle Eastern dress.

DR JANET STARKEY is a Lecturer at the Centre for Middle Eastern and Islamic Studies at the University of Durham. She has written and edited numerous publications, including *Desert Travellers: From Herodotus to T.E. Lawrence* (ASTENE 2000), *Interpreting the Orient*, (Ithaca 2001), *Travellers in Egypt* (I.B. Taurus 1998 and 2001) and *Unfolding the Orient and Gold* (Ithaca 2001).

PROFESSOR MALCOLM WAGSTAFF is Professor Emeritus at the University of Southampton and Visiting Fellow in its Department of Geography, where he served for many years as a specialist in historical geography and the Middle East. He retired from the chair of the executive committee of ASTENE in 2005. He has travelled extensively in Greece, Turkey, Cyprus, Syria and Israel/Palestine. His publications include *The Evolution of Middle Eastern Landscape: An Outline to AD 1840* (Croom Helm 1985), and *Greece: Ethnicity and Sovereignty, 1820-1994: Atlas and Documents* (Archive Editions 2002). A biography of Colonel Leake (1777-1860) is in progress.

# Publisher's Note

This eclectic work carries the reader from the political travels of the Cicero brothers, amid the further reaches of the Roman Empire, to the missionary exploits of a Chaldean Christian in Iraq eighteen centuries later; from the adventures of a French dragoman in Ottoman Egypt to the researches of Scottish Enlightenment naturalists in Aleppo. There are examples in this volume of a wonderful variety of Near Eastern travel, by a catholic collection of travellers.

The term 'Near East' is interpreted loosely here. We embrace Egypt and Greece when such an embrace is called for. Thus do we find an early Irish photographer and his wife among the temples of Abu Simbel and Kom Ombo. We follow French mapmakers as they survey the Morea, and accompany an amateur Egyptologist wrestling to interpret the hieroglyphics at Alexandria. We also hear of some of the 19th century travellers who braved 'tribesmen and troubles' in their quest to see the rose-pink city of Petra. We join a Scottish officer in his excavations at Mausolus' tomb in Halicarnassus, we follow a dapper French aristocrat on his artistic sojourn in the Levant, and accompany a Presbyterian cleric on a quest to track Biblical clues in the landscape of the Holy Land. The collection ends with an investigation, relevant to the concerns of our epoch, into the politics and peregrinations of Gertrude Bell in Iraq a century ago.

This volume of essays has been assimilated from the proceedings of the 2001 Edinburgh Conference of the Association for the Study of Travel in Egypt and the Near East (ASTENE), an organisation which brings together all those interested in the history of travel, in a region of the world that has captivated the imagination of Europe from ancient times.

This new, revised printing of *Travellers in the Near East* incorporates a number of amendments and corrections to the 2004 edition.

## CHAPTER ONE
# A Tale of Two Ciceros: Travels in Asia Minor in the late Roman Republic

## Marsha B. McCoy

The brothers Cicero – Marcus the famous politician and his younger brother Quintus – both served as governors in Roman provinces in Asia Minor; Quintus first in the province of Asia in 61-59 BC, and Marcus later in Cilicia in 51-50 BC. Although just a decade apart, their experiences were very different. These differences were partly due to the changing fortunes of the Ciceros during a turbulent decade, but even more to the very different provinces they went to. Though geographically close, the history, culture and terrain of Asia and Cilicia were miles apart, as were their relationships with Rome. And Rome's treatment of these provinces changed markedly over the decade, reflecting the shift of her concerns as she approached the civil war that was to bring down the Republic. The letters between the Cicero brothers and to their friends, particularly to Cicero's close friend Atticus, give us a unique insight into the Roman experience in the Near East and show us how, then as now, foreign policy changed with internal political affairs.

Rome had no provinces until the end of the Second Punic War in 241 BC, when she annexed Sicily from Carthage and formed her first overseas territorial possession. Corsica and Sardinia soon followed. At first Rome increased the number of praetors – the second-highest office in the Roman state after the consul – to handle the administration of these new territories and the deployment of troops. But then Rome moved to a new system. Instead of creating more new praetors as provinces increased, or developing a new class of officials to govern these provinces, the Romans began appointing men who had previously served as praetors or consuls to govern in place of current praetors or consuls. These officials therefore became known as propraetors or proconsuls, and this is the system that the Romans used henceforth to administer their growing empire.[1]

The territory that became the province of Asia – basically the westernmost portion of modern-day Turkey – was left to the Romans in 133 BC by its last independent ruler in his will, and the Romans made it a province shortly thereafter. Attalus had bequeathed to the Romans a kingdom rich in fertile river valleys that produced a variety of agricultural products. Asia was blessed with numerous harbours that provided the routes for goods from the interior and beyond out to the Mediterranean and countries west. And, finally, Asia boasted some of the wealthiest, most beautiful cities in the ancient world, cities like Ephesus, Pergamon and Aphrodisias, with their fine temples, theatres, and municipal buildings. It was to this province, with its distinguished cultural history, that Quintus Cicero was sent in early 61 BC.[2]

This period was a time of relative political stability in Rome. The conspiracy of Catiline, which had marked the second half of Marcus Cicero's consulship in 63 BC, had ended with the death of Catiline in battle in early 62 BC. December 62 BC saw the Bona Dea scandal, in which the nobleman Clodius infiltrated a female-only religious ritual dressed up in women's clothing. Much to the shock of Cicero, who testified against him at the trial in mid-61, Clodius was acquitted thanks to lavish bribery of the jury.[3] The year 60 witnessed an informal political alliance of Caesar, Pompey, and Crassus, the most powerful men of the late Republic. Caesar became consul in 59 BC and in the same year Pompey married Caesar's daughter Julia, strengthening the ties between the two men.

For the Ciceros, too, these were good years. They were at the height of their power and prestige. Marcus Cicero had served as praetor in 66 and consul in 63, both at the minimum age. His younger brother Quintus had benefitted from Marcus' success too. Quintus was elected aedile for 65 at the elections of 66, when Marcus was praetor. And he was elected praetor for 62 at the elections of 63, when Marcus was consul. The Ciceros were a new family – one whose members

*Figure 1:* Cistophorus *('cistophoric' tetradrachm, the normal silver currency of Asia Minor, so called because it usually had a representation of the sacred cists or chest of Dionysus between guardian snakes, also attributes of Dionysus, who was traditionally from Phrygia). It was struck at Apamea in Phrygia, part of the province of Cilicia 56-50 BC, during Marcus Cicero's governorship. It carries the legend on the top of the coin in Latin: CICERO M.F. PROCOS (Cicero, son of Marcus, proconsul). The legend on the bottom of the coin is in Greek.*

had not held political office in Rome before. Indeed, the Ciceros were not even from Rome – they came from Arpinum, a town south-east of Rome in the foothills of the Apennines. Quintus' governorship of Asia was the first either of the Ciceros had held. Marcus had spent the two years after his praetorship campaigning almost constantly for the consulship. In early 62, after his year as consul, he had traded away the consular provinces of Cisalpine Gaul and Macedonia to others in order to gain political favour from them. So Marcus saw Quintus' time in Asia as an opportunity to display the Ciceros' philhellenism, and to garner further prestige and acclaim among the Roman elite by winning the provincials' respect with Quintus' fair and judicious administration of the province.[4]

Marcus wanted to make several political points. In 70 BC he had successfully prosecuted the Roman governor of Sicily on behalf of the Sicilians for blatant and devastating extortion in the province. In this case Marcus had established himself as the pre-eminent lawyer of his age, but he had also earned a number of important enemies. So it was very important to Marcus that Quintus vindicate himself beyond the shadow of a doubt as an incorruptible governor and administrator. The province of Asia presented an ideal territory for such vindication because Asia's wealth, like Sicily's, had long proved an irresistible temptation for propraetors and proconsuls bankrupted by expensive political campaigns in Rome. Thus Marcus wrote to Quintus in one of the long letters he sent to him while his brother was in Asia:

> Therefore, since so great theatre has been given for your virtues to display themselves, the whole of Asia no less, a theatre so crowded, so vast, so expertly critical, and with acoustic properties so powerful that cries and demonstrations echo as far as Rome, pray strive with all your might not only that you may appear worthy of what was achieved here but that men may rate your performance

above anything that has been seen out there. As chance has decreed, my public work in office has been done in Rome, yours in a province. If my part stands second to none, make yours surpass the rest.[5]

Marcus worked to extend Quintus' tenure in Asia for a second year, as he said to Quintus: 'thinking of the welfare of the provincials, opposing the effrontery of certain businessmen, and seeking to add to our prestige by your abilities.'[6] And when Quintus' governorship was extended for a third year Marcus wrote:

...like good playwrights and hard-working actors, take your greatest pains in the final phase, the rounding off of your appointed task. Let this third year of your term as governor be like the last act of a play – the most highly finished, best fitted-out of the three.[7]

Elsewhere in these long letters to Quintus, Marcus gave his brother detailed advice on how to behave and how to handle cases of provincial justice that came before him in his capacity as arbiter of Roman law. And he helped ensure that reports of Quintus' popularity in Asia reached the right ears in Rome.[8]

But the decade of the 50s saw major shifts in both the situation in Rome and the fortunes of the Ciceros. In 59 Caesar, then both consul and Pontifex Maximus, or head priest of Rome, permitted the transfer of the patrician nobleman Clodius into the plebeian family of that name, and Clodius thereupon stood for election to the plebeian tribunate in December of that year. The plebeian tribunate was an office that had been used for decades to oppose the views and interests of the Senate, and bills proposed by plebeian tribunes and passed in plebeian assemblies had the force of law.

One of Clodius' first acts in 58 was to propose a bill outlawing anyone who had executed citizens without a trial. The bill was clearly aimed at Marcus Cicero, who as consul in 63 had authorised the dubious execution of several of Catiline's co-conspirators. Clodius obviously meant to punish Marcus for his damaging testimony against him three years before at the Bona Dea trial. Thus in early March 58, Cicero did not find himself basking in the reflected glow of Quintus' fine administration and his own élite status as a former consul. Instead, he was fleeing Rome for his life through southern Italy towards Brundisium, from where he sailed to Greece to await news of his fate.

The Senate recalled Marcus in 57 BC once Clodius was out of office. Marcus returned to Rome to great acclaim on 4 September 57, but he never regained his former footing. Already in early 59 he had attempted, without success, to block the extension of Quintus' governorship for a third year. Had he been able to do that, Quintus would have been back in Rome by the summer of 59, having achieved in his two year term in Asia all that the Ciceros wanted, and he would have been able to help Marcus in the following year against the attacks of Clodius. As it was, Quintus returned to Rome from Asia just as Marcus was fleeing Rome for Greece, and they missed seeing each other in the fear and haste of the moment. Such was the momentum that Clodius seemed to have built against Marcus, that he feared Quintus might even be prosecuted on his return to Rome for corruption while governor.[9] After Marcus was recalled to Rome in 57, he participated far less in public life, and it is to this unsought and reluctant retirement that we owe many of his great philosophical works, notably *De Oratore* (55), *De Republica* (54), and *De Legibus* (52).

The 50s in Rome witnessed a similar deterioration in the political stability of the state. In 58, after his year as consul, Caesar received a five-year proconsulship in Gaul, ostensibly to pacify the unruly tribes to the north. He began building an army superior to any other in the Roman world. As Caesar became stronger militarily in Gaul, and Pompey and Crassus became stronger politically in Rome, they decided to revive the power-sharing arrangement worked out in 60. In 55 Caesar's command in Gaul was extended for another five years, and Pompey and Crassus became consuls for a second time. This, of course, had the effect of strengthening them further in their respective spheres and raising the stakes of their political manoeuvering. Then in 54 Julia, daughter of Caesar and wife of Pompey, died. In 53, Crassus and his son died fighting the Parthians at Carrhae in Mesopotamia, and Pompey married the younger Crassus' widow in order to bring Crassus' supporters over to his side. The consular elections in 53 for 52 could not be held because of civil unrest in Rome. Finally, in 52, Pompey was appointed sole consul, and began working to restore order in the state.

One of the new laws that Pompey promulgated required Roman officials to wait five years after the end of their tenure in office before serving as governors in provinces. This was intended in the long run to reduce corruption in provincial administration by preventing Roman politicians from spending their way into poverty on political campaigns, and then recouping their losses in the following year by squeezing the provincials. The immediate effect of this law, however, was to press into provincial service any former consuls or praetors who had not previously served as governors, since recent and current office-holders now had to wait five years before assuming governorships abroad. And so Marcus Cicero, a confirmed stay-at-home *bon vivant* and aesthete, in the middle of writing *De Legibus,* suddenly found the province of Cilicia thrust upon him in 51 BC.[10]

There could hardly have been a worse mismatch of province and governor. Unlike Asia, with its Greek cities, its wealth and culture, its pleasing landscapes, and its long relationship with Rome, Cilicia was out of the Greek sphere. It was rough, uncultured, a traditional haven for pirates, and had only been annexed as a province in 66 BC. Furthermore,

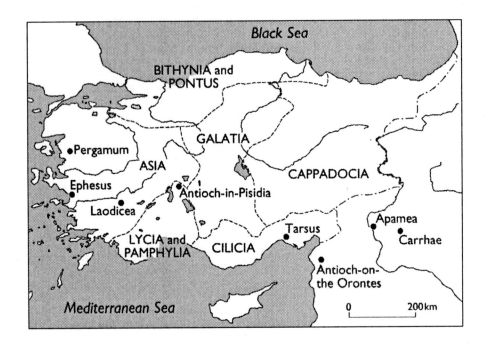

*Figure 2: Map of Asia Minor in the time of the Ciceros.*

with the death of Crassus at the hands of the Parthians, Cilicia was threatened from the East by these same Parthians, since it provided the only route from Syria to eastern Asia Minor (through Synnada, Apamea, and Laodicea to Ephesus). Crassus, the richest man in Rome, had hoped with a victory over the Parthians, to attain some of the military glory enjoyed by Pompey, and more recently Caesar. With his death and the resulting civil turmoil in Rome, interest in Rome's easternmost territories had lapsed. Marcus, with no military experience since his youth, persuaded Quintus to accompany him as legate. In addition to Quintus' experience as governor in Asia, he had also served as legate to Pompey in Sardinia in 57 and 56, and had just returned in early 51 from three years of distinguished service with Caesar in Gaul.[11]

Marcus and Quintus arrived in Cilicia in mid-51 BC to find two underpaid, understaffed legions, and Crassus' embattled quaestor holding the fort in Syria to the east. A Parthian raid on Antioch, the capital of Syria, was rebuffed, but Marcus feared the Parthians might attack the highlands of Cappadocia, a client-kingdom of Rome, or cross the Amanus mountains dividing Syria from Cilicia. He enlisted the help of Galatia, another client-kingdom to the north, and in two successful campaigns against the Parthians in the autumn of 51, first extracted the ceremonial acclamation of Imperator from his soldiers, and then earned a *supplicatio* or 'festival of thanksgiving' from the Senate in Rome. Marcus returned to Italy in the winter of 50, but remained outside the city of Rome so that he could retain his promagisterial power to command troops. Clearly he hoped that the Senate would award him a triumph through the streets of Rome before he had to relinquish his *imperium*. But events outstripped his hopes and ambitions. As we all know, Caesar defied the Senate, crossed the Rubicon River in north-eastern Italy in early 49, defeated Pompey at the Battle of Pharsalus in 48, and was himself assassinated on the Ides of March, 44. Marcus and Quintus Cicero met their deaths in 43 at the hands of the henchmen of Antony, who had been Caesar's lieutenant and took over his faction after Caesar's assassination.[12]

So Marcus never received his triumph, and his and Quintus' years in Asia Minor became short episodes in lives generally spent at the centre of power and at the forefront of the rhetorical and political thought of the age. Asia remained a province well into the late Roman Empire, but Cilicia was dissolved not long after Marcus' governorship there, and its territories were distributed between Syria to the east and Galatia to the north.

History has a way of commemorating itself in unexpected ways. Marcus and Quintus Cicero's writings and letters have survived to recount for us the dramas, high and low, of the late Roman Republic. There also survives the coin depicted in the illustration (Fig.1). This fine example of a *cistophorus*, the standard currency used in Asia Minor in this period, was clearly made by local minters especially to commemorate and honor Cicero's year as governor of their province.[13] The dual combination of the Latin and Greek legends made it readable both to local Greek-speakers and to Roman officials and businessmen. Its visual and verbal language evokes, perhaps as much as do Marcus Cicero's many writings, the memory that he would have wanted us to have of him: a man of works, a man of letters, and a cultured philhellene who served his country, its citizens and its subjects honourably and well.

## REFERENCES

1. Broughton, T.R.S., *The Magistrates of the Roman Republic*, 3 vols, New York: American Philological Association, 1951-52; Mommsen, T., *The Provinces of the Roman Empire*, Chicago: University of Chicago Press, 1968.
2. Jones, A.H.M., *The Cities of the Eastern Roman Provinces*, 1971; Magie, David, *The Roman Rule in Asia Minor*, 1950.
3. Shelton, Jo-Ann, *As the Romans Did: A Sourcebook in Roman Social History*, Oxford University Press, 1988.
4. Cicero, *Letters to Atticus*, ed and trans. D.R. Shackleton Bailey, Loeb Classical Library 1999, Letter 15, 1.
5. Cicero, Letters to his Brother Quintus, ed. and trans. D.R. Shackleton Bailey, Loeb Classical Library, 2002, Letter 1, 42-43.
6. ibid, 2.
7. ibid, 46.
8. Magie, 1950.
9. Cicero, *Atticus*, Letter 54, 1.
10. Lintott, A.W., *Imperium Romanum: Politics and Administration*, London, New York: Routledge, 1993.

11. Sherwin-White, A.N., *Roman Foreign Policy in the East*, University of Oklahoma Press, 1984.
12. Wiedemann T., *Cicero and the End of the Roman Republic*, London: Bristol Classical Press, 1994.
13. Sutherland, C.H.V., *Roman Coins*, New York, 1974.


BIBLIOGRAPHY

Cicero, *Letters to Atticus*, ed. and trans. D.R. Shackleton Bailey, Cambridge, MA , London: Loeb Classical Library, 1999.
—, *Letters to his Brother Quintus*, ed. and trans. D.R. Shackleton Bailey, Cambridge, MA, London: Loeb Classical Library, 2002.

Broughton, T.R.S., *The Magistrates of the Roman Republic*. 3 vols. New York: American Philological Association, 1951-52.

Jones, A.H.M., *The Cities of the Eastern Roman Provinces*, 2nd ed. Oxford: Clarendon Press, 1971.

Lintott, A.W., *Imperium Romanum: Politics and Administration*. London, New York: Routledge, 1993.

Magie, David, *The Roman Rule in Asia Minor*, Princeton: Princeton University Press, 1950.

Mommsen, T., *The Provinces of the Roman Empire*, Chicago: University of Chicago Press, 1968.

Shelton, Jo-Ann, *As the Romans Did. A Sourcebook in Roman Social History*, Oxford: Oxford University Press, 1988.

Sherwin-White, A.N., *Roman Foreign Policy in the East, 168 BC to AD 1*, Norman, OK: University of Oklahoma Press, 1984.

Sutherland, C.H.V., *Roman Coins*, New York: G.P. Putnam's Sons, 1974.

Wiedemann, T., *Cicero and the End of the Roman Republic*, London: Bristol Classical Press, 1994.

# Chapter Two
## *Mercantile Gentlemen and Inquisitive Travellers: constructing* The Natural History of Aleppo

### Janet Starkey

> The mosques, the minarets, and numerous cupolas form a splendid spectacle, and the flat roofs of the houses which are situated on the hills, rising one behind another, present a succession of hanging terraces, interspersed with cypress and poplar trees.[1]

Introduction

So much has been written about the rise of Romantic Orientalism in the 19th century that it seems important to explore earlier influences on those of European travellers, such as the physicians Alexander and Patrick Russell who lived in Aleppo in the mid-18th century, to understand more of the history of East-West contact in the Levant. *The Natural History of Aleppo* (1756; second edition 1794) is more than simply a traveller's account of the city in the 18th century. The book also influenced the mid-19th-century ethnographer and lexicographer, Edward W.

Lane (1801-76), whom Edward Said associated with the rise of Romantic Orientalism. Said erroneously talks of Dr Russell's account as 'a forgotten work'.[2] According to Sari J. Nasir,[3] in his Preface to *An Account of the Manners and Customs of the Modern Egyptians,* Lane mentions that the English reader's knowledge of the Middle East at the end of the 18th century was primarily based on *Aleppo.* Lane certainly continued in his own work the Russells' fine tradition of exquisite recording and providing detailed descriptions of the country and social customs. The Russells not only made their own observations, but provided comparative detail from classical and Arabic texts as well as from British and European travel accounts. They also attempted to keep abreast of the times: the use of Linnaean terminology for botanical specimens is evidence enough.

Alexander Russell (c.1715-68) published *The Natural History of Aleppo* in 1756, a book that he began shortly after his arrival in Aleppo in 1740. After Alexander Russell's death in 1768, the second edition of *The Natural History of Aleppo and parts adjacent...*, was revised and enlarged by his half-brother, Patrick Russell, MD (1726-1805) and published in two volumes, with wonderful illustrations, in London in 1794,[4] with many additional notes and references to earlier travellers in the Levant and to Arabic sources relating to the city and medical practice. The 1794 edition of *Aleppo* also includes paintings by the famous Anglo-Egyptian flower painter George D. Ehret, a colleague of Alexander's friend, Dr John Fothergill.[5] Preparation of the second edition was not without stress for Patrick. He wrote: 'The death of the Author, in 1768, caused a temporary interruption of studies, which his Brother found himself unable to resume, without suffering, by association, many painful recollections, which for a long while, too sensibly perhaps, affected his mind.'[6]

## The Russells: European Outlook and the Scottish Enlightenment

Alexander Russell, MD, FRS, third son (by his second wife) of a lawyer, John Russell, was born about 1715 in Braidshaw, Midlothian and was educated at Edinburgh High School and later at the University of Edinburgh. Alexander Russell worked as the physician to the Levant Company in Aleppo from 1749 to 1753 and then returned to London in 1755. Alexander's half-brother, Patrick Russell, MD, FRS, was born in Edinburgh on 6 February 1726/7. Educated possibly at King's College, Aberdeen, he gained his M.D. at Edinburgh University. Arriving in Aleppo in 1750, Patrick was also a physician with the Levant Company from 1753, a position that he held until 1771.

The Russells went to university at a time when the city of Edinburgh was flourishing as a centre of the arts and philosophical thought at the beginning of the Scottish Enlightenment. The classic period of the Scottish Enlightenment, from the early 1740s to the late 1790s, witnessed many scientific and pioneering publications, together with the founding of philosophical clubs and societies in the major Scottish cities. Edinburgh was a centre of excellence in a wide range of fields including medicine, geology, agriculture, natural history, philosophy, poetry and painting.

The development of a high standard of education within the five important universities in Scotland during this period was enhanced by vigorous dialogue on ideas generated by a small group of thinkers, who possessed originality and even genius. At the time that Alexander Russell went to the University of Edinburgh, the average age of students was between 14 and 16 years old. Many university students were from the middle ranks of society – the very group of people who exported the ideas of the

Enlightenment through commercial ties and developed colonial infrastructures in India, America and Africa. It is against this background that we can place Alexander Russell during his medical training at the University of Edinburgh between 1732 and 1734.

Despite the Union with England in 1707, Scotland in the 1730s to 1750s was more in tune with continental attitudes than those of London: Scots were educated in both Scottish and European traditions. There were significant links in Scotland with the French Enlightenment in Paris and strong connections with Leyden, especially in the field of medicine. Scottish scholars went to European universities and then returned to teach in Scottish universities. Almost all Alexander's medical lecturers, including the five members of the College of Physicians, had studied in Leyden with the Dutch physician and Professor of Medicine, Herman Boerhaave (1668-1738), or in other centres including Rheims, Genoa and Paris. Boerhaave is credited with founding the modern system of teaching medical students by using a clinical or 'bedside' manner rather than simply relying on classical textbooks. He was renowned as a brilliant teacher, and students from all over Europe came to hear his lectures. Through his pupils, he exerted great influence on later medical teaching in Edinburgh, Vienna and Germany. These included Alexander Russell's teachers: Alexander Monro *primus* (1697-1767), the founder (in 1720) of the Edinburgh Medical School and the Professor of Anatomy there; John Innes (d. 11 Jan 1778), the Professor of Medicine with John Rutherford (1695-1779) – (it was Rutherford who promoted the establishment of a botanical garden in the city of Edinburgh); the botanist Professor of *Materia Medica*, Charles Alston (1683-1760); Dr Sinclair, who lectured on the Theory of Physic; and the Professor of Chemistry, Dr Plummer. Boerhaave's approach was to encourage students to seek answers to problems for themselves. His influence is reflected in the Russells' approach to their account of life in Aleppo.

Neither Patrick nor Alexander worked alone in Aleppo, for they had maintained friendships formed during their university and medical careers. These friendships were to have a profound influence on the creation and format of *Aleppo*. William Cuming (1714-88), George Cleghorn (1716-89) and Alexander Russell founded a small but active student medical society at the University of Edinburgh in 1734 that was to become the Royal Society of Edinburgh.[7] Alexander's fellow student, John Fothergill, the famous Quaker medical practitioner and botanist, who had a long and distinguished career in London (and was a friend of many notable figures, including Benjamin Franklin in the States and Carolus Linnaeus in Upsala), joined the society in his second year, 1735. About 400 letters written by Fothergill have been found. They reflect his devotion to this group of fellow students. It was Fothergill who urged Alexander Russell to put the results of his studies in Aleppo into a book after Russell returned to London in 1755. Fothergill was even present at Alexander's deathbed in London 1768.

Cuming, the son of a prominent Edinburgh merchant, studied with Boerhaave in Leyden from 1735.[8] Cleghorn went on to practise in Minorca in 1736 for 13 years as an army surgeon with the 22nd Regiment of Foot. Fothergill kept him supplied with books and with letters filled with information about the latest scientific trends.[9] Cleghorn regularly sent botanical specimens back to Fothergill and, at the latter's instigation, published *Observations on Epidemical Diseases of Minorca* (1751).[10] Like Russell, Fothergill and Cuming, Cleghorn was thoroughly competent in Latin and quoted copiously from classical sources. In many ways, *Aleppo* conformed to the model set out by Cleghorn, for both books contain much medical detail, with observations on the climate, vegetable and animal life, and the conditions of life of the inhabitants. As Fothergill mentioned, both Cleghorn and Russell 'have acquired much reputation, and have done their country most signal service' by their publications.[11]

*Writing* The Natural History of Aleppo

Russell published *Aleppo* the year after his return to London. In the same year he was elected as Fellow of the Royal Society. He also contributed several papers to its *Transactions*.[12] The quarto monograph, with its large print, illustrations, footnotes and wide margins, was an instant success and a Dutch translation was published in 1762. In 1782, selections from *Aleppo* were translated into Arabic.[13]

There were also several contemporary abbreviated versions including *A Description of Aleppo, and the Adjacent Parts*. This was published in the second volume of *A Compendium of the Most Approved Modern Travels* in 1757,[14] with excerpts from other travellers, including Dr Richard Pococke, who travelled in the East from 1737 to 1740; Alexander Drummond, who first visited Aleppo in 1745 and later became its Consul 1754-6,[15] and to whom the first edition was dedicated, as well as to the 'Gentlemen of the British Factory at Aleppo; and those now in England, who have formerly resided there';[16] and Revd Henry Maundrell, with a section from his *A Journey from Aleppo to Jerusalem at Easter*, AD *1697*.[17] Another edition of Russell's abridged version of 1757 appeared in *The World Displayed* (1779).[18] Other extracts in this 1779 volume are from Dr Richard Pococke's *Travels through Egypt* and his *Journey to Palmyra*. The book also contains 'The Travels of the Ambassadors from the Duke of Holstein into Moscovy, Tartary, and Persia'. The last chapter of the 1779 version is a summary of text in *Aleppo* (1756), with associated plates.[19]

The first volume of the 1794 edition contains a description of the city and its inhabitants. Topics in volume II include natural history, descriptions of Syria and Aleppo, with much detail on domestic culture, manners and customs, monuments and common diseases, including a vivid description of the effects of smallpox. The book has been described by Sarah Searight as 'a delightful and exhaustive survey of the society, flora, fauna and particularly the plague' (on which Patrick became an expert) of a major Ottoman city.[20] It is a better known and better written example of erudite travelogue feeding 'somewhat indigestibly the [West's] growing curiosity in the eastern Mediterranean, the Levant and Egypt'.[21]

The Russells, like John Fothergill, William Pitcairn and many other of their Edinburgh colleagues, were fascinated by the natural world. The 1756 edition, third chapter of the 1757 version and Volume II of the 1794 edition are full of detail about birds, animals (domestic and wild), trees and flowers:

> The country also produces several kinds of forest trees, as the plane, the white poplar, the horn-beam, a very few oaks, the ash, the tamarisk, the turpentine tree, and many others.
>
> There are here likewise a great variety of garden plants and flowers, which render the country extremely pleasant in spring, before the great heats have scorched them up, and after the succeeding rains have revived their beauties.[22]

They also describe the Aleppine gardens that provided most of the seasonal fruit and vegetables for the town:

> Of the fruits of this country, there are only two or three sorts of apples, and those very bad. They have cherries, apricots, peaches; indifferent good pears, quinces, pomegranates of three sorts, mulberries, oranges, lemons, figs of four kinds, walnuts, hazle-nuts, pistachio nuts &c. These trees are all standards planted promiscuously, and little improved by culture.[23]

Alexander Russell spent a lot of time constructing

taxonomic systems, and included plants, mammals, reptiles and insects in an attempt to 'arrange nature' in the 1756 edition. This edition, however, was written before Carolus Linnaeus (1707-78) established the binomial classification of plants. Sir Joseph Banks[24] and Daniel Carl Solander (1736-82), a pupil of Linnaeus who went to assist Joseph Banks at the British Museum in 1759, helped to rework the botanical terminology in *Aleppo*, as Patrick acknowledged:

> The Catalogue of Plants, growing in the vicinity of Aleppo will be found to have undergone material alteration, and to be much improved. But it is my duty to acknowledge that this is to be ascribed to the friendly assistance of Sir Joseph Banks, (and the late Doctor Solander), who, with their usual readiness to countenance every attempt tending to the advancement of Natural History, bestowed many hours on the examination of a large Collection of Specimens from Syria; and, after correcting numberless errors in the former arrangement, composed the classical catalogue now substituted for the old one.[25]

The brothers shared an interest in natural history with other ex-student friends, medical colleagues and professors. Alexander sent seeds of an elegant shrub, the true scammony plant from Aleppo, to the nurseryman James Gordon and to the Quaker and botanist Peter Collison FRS (1693-1767) who had a garden at Mill Hill from 1749. In 1754 he also sent seeds of *Arbutus andrachne* (a strawberry tree), collected from the mountains above Aleppo, to be propagated by James Gordon in his nursery in Essex Road, Mile End and by Fothergill.[26] No doubt both Russells sent other samples back to their friends, and when in London would have visited Fothergill's botanical gardens in Upton Park. William Pitcairn (1711-91), who was present by Alexander's deathbed, also maintained a botanical garden in Islington 'second only in size and importance to Dr Fothergill's at Upton'.[27]

## The Russells in Aleppo

The city of Aleppo was founded in the 14th century BC and was closely associated with Saladin (d. AD 1193). Patrick Russell used contemporary Arabic sources, sometimes in Latin translation, to describe the history of the city in *Aleppo*.[28] The town was incorporated into the Ottoman Empire in 1516 and remained prosperous until the end of the 18th century. From 1517, Aleppo was the chief town of a Turkish *vilayat* in the Ottoman Empire, later ruled by Sultan Mahmud I (1730-54), Osman III (1754-7) and Mustafa III (1757-74). It must be remembered that Ottoman expansion into Europe continued as late as 1687, at Mohacs, and that the Ottoman Empire was still a great military power until the 1760s. Charles Perry MD (1698-1780), a traveller in the region in 1739 and 1740, wrote in *View of the Levant*,[29] of the 'present weak, feeble condition of the Turkish Empire' and of the idyllic life of traders in Aleppo.[30] Alexander Russell became a friend of and adviser to the Pasha in Aleppo and his kindly reputation spread throughout the Ottoman Empire.

Aleppo was an important trading centre on the road between Asia and Europe, which covered the 80 miles of desert and mountain to the Mediterranean ports of Lattakia (al-Ladhiqiyya), Tripoli, and Scanderoon (Iskenderun). The vitality of East-West mercantile contacts based in Aleppo were stimulated by its position as entrepôt, despite the unreliable shipping schedules in the Mediterranean. Aleppo was an emporium through which passed a constant stream of traders and officials from the East India Company, especially in the mid-18th century, for the city commanded the 'Great Desert Route', the shortest overland route to India and Persia,[31] via the Euphrates and along caravan routes to Mosul, Baghdad and Basra in Iraq, a distance of about 760 miles. Other regions in Aleppo's hinterland included southern Anatolia, and Smyrna (Izmir).

Government officials, military personnel, pilgrims to Mecca and Jerusalem (although Patrick Russell noted that Aleppo's trade with Mecca and Arabia declined in the late 18th century),[32] as well as local villagers in the Pashalik, and Arabian and Syrian desert Bedouin camel-drivers and herders of sheep and goats, all visited the city. The most important exchange was English woollen cloth for Persian, Syrian and Ottoman silk fabrics, and Indian spices.[33]

According to Russell's figures, in the 18th century Aleppo was the third largest metropolis in the region: 'The inhabitants of the city and suburbs of Aleppo are computed at about 235,000, of whom 200,000 are Turks, 30,000 are Christians, and the remaining 5,000 Jews'.[34] As Alexander Russell noted:

> Tho' they are of such different religions, they seem to be nearly the same people: nor are the Christians much superior to their neighbours in virtue. The greatest number of them are Greeks, the most numerous next to them are the Armenians, next to them the Syrians, and then the Maronites; each of whom have a church in a part of the suburbs, where most of them reside. The vulgar language is Arabic; but the Turks of rank use the Turkish: most of the Armenians can speak Armenian; many of the Jews understand Hebrew; but few Syrians can speak Syriac; and scarce one of the Greeks understands a word of either ancient or modern Greek.[35]

The Russells also detailed the exciting variety of ethnic groups, the domestic manners of the inhabitants of the city and Aleppo's gardens, coffeehouses, its government and commerce. The accounts reveal the Halabis' obsessive attention to the complexities of their elaborate dress codes: silk and furs, symbols of power, elegance and luxury, dominated upper-class culture (Fig. 4).[36] The Russells noted the Halabi preoccupation with elaborate rules of etiquette and ceremony that dictated proper behaviour in different situations.

The Russells also witnessed emigration after religious persecution of Aleppine Christians after 1730, periodic economic slumps, and official oppression. Others migrated to trade in the major cities of Egypt, Iraq, Syria and Anatolia; Christian and Jews went to Livorno and India, whilst scholars were attracted to Istanbul.[37] They noted that 1756 was a very bad winter with heavy snow and loss of life, resulting crop failure and the danger of famine.[38] By 1757 bread was scarce and the governor sold his supplies at excessive prices. In the winter of 1757-8 hunger was widespread and many were unemployed. As a result, epidemic diseases affected the malnourished, and as many as 40,000 died of starvation, cold and disease. There were further food shortages in 1761 and between 1764 and 1768.[39] By the 1770s Patrick Russell noted many ruined, abandoned settlements around Aleppo: 'The Olive Tree Village and others are totally deserted. It is asserted that of three hundred villages, formerly comprehended in the Bashawlick, less than one third are now (1772) inhabited: agriculture declines in proportion.'[40]

Alexander had witnessed outbreaks of the plague in 1742-4 and saw the economy of Syria collapse in the 1750s. During his residence in Aleppo, Patrick experienced the food shortages of 1751 and the plague in 1760-2, which reduced the population to around 150,000 people. Alexander Russell was consulted by the Privy Council about a threatened outbreak of the plague in England[41] and in 1791 Patrick Russell published *A Treatise of the Plague*[42] with case studies of communicable diseases, especially the bubonic plague in 18th-century Aleppo, and dealing with Aleppo's hospitals and quarantine procedures.[43] One can speculate on the association of the spread of the plague with the movement of large caravans from the East. He gave a detailed breakdown of mortality estimates of 7,767 deaths in 1761 and 11,883 in 1762 – around a 15-20 per cent mortality rate in the city. He made many medical observations and public responses to the disease, and established himself as a leading authority on the plague.[44]

In addition to references to plagues in *Aleppo*,[45] there is a detailed description of local diseases and epidemics recorded between 1742 and 1753,[46] including a discussion of the 'Aleppo boil', and notes on infant morality.[47] In December 1768 Alexander and Patrick published an account of inoculation in Arabia; a note of correspondence between the two brothers in which they recount how Alexander was informed that inoculation against smallpox was practised by Bedouin.[48] Patrick also wrote on earthquakes in Syria,[49] and both were interested in climatic effects on health. There was a high rate of sickness among Europeans who suffered from dysentery, malaria, and other endemic diseases.[50]

*'Blending matters collected from reading'*[51]

From the late 16th to the end of the 18th century, the occasional European traveller visited Aleppo, resulting in as many as 23 published accounts of the city.[52] What is particularly of interest is the use that Patrick made of travellers' accounts. He recognised that: 'Since the beginning of the 17th Century, the Curious in Europe owe most of what they have learned relating to modern Syria, either to the casual remarks of mercantile Gentlemen settled abroad, or to the researches of a few more inquisitive travellers.'[53] Accounts were often published within a relatively short period of their return to Europe, and many of these contain precise records of many details with associated drawings and sketches. In addition, their use of appropriate Arabic and European texts reflects a serious scholarship. One can also speculate that the Russells had access to the medical library of their friend Dr Fothergill, a library that included Latin editions of classical, mediaeval and Renaissance medical texts, including Pliny the Elder, Dioscorides, Galen, Aristotle, the medical classics of Arabic, and Buffon's *Histoire Naturelle*, many of which are quoted by Patrick Russell in the notes and text of *Aleppo*.[54]

Patrick describes the creation of the second edition in his introduction:

> For many years before he engaged in the present Work, he had little leisure for perusing the journals of Eastern travellers; and after his return to Britain, he resolved, with a view to avoid blending matters collected from reading, with what might be suggested by his experience in Turkey, not to look into Books of Travels, till he should have sketched from recollection, all he meant to insert as supplementary to his Brother's Book.[55]

Patrick subsequently explored earlier classical and Arabic texts and accounts from early European travellers with what we would still consider to be a thoroughly modern academic approach. The range of sources they used is impressive: 'In this course of reading, some of the early travels were perused with much satisfaction.'[56]

*Classical Sources*

The Russells received their medical education in Edinburgh from lectures given in English or Latin, so they had no difficulty in using classical accounts of the region, as well as classical sources, in their medical practice. Just as Alexander's university friend, Cleghorn, used Hippocrates and Celsus in his book on Minorca, the Russells included classical sources such as Aeschylus,[57] Juvenal on blacking eyelids, for example,[58] Dioscorides, Naumachius and Pliny the Elder, for example, on khol,[59] and Homer's *Odyssey*.[60] He also cites Galen (d. c. AD 216) the Greek physician and medical writer, whose 150 or more medical works were assiduously translated into Arabic. Patrick used Latin translations of his works which were published in Basel in 1538 and Venice in 1625.[61] Galen's work was the authoritative standard for medical knowledge until the 19th century and was considered to be essential reading for any doctor; it was practically impossible to hold a view contrary to Galen in 18th-century Aleppo.

*Arabic Sources*

The Russells mention the professional storytellers and poets who would often recount tales from the *Arabian Nights*, entertaining crowds with a repertoire of exotic stories full of genies, magic, supernatural birds and erotic veiled dancing girls. Storytellers sometimes broke off in the middle of their tales leaving listeners to speculate on the outcome until next day.[62] These tales were also retold by women in the privacy of their bedrooms, to send their menfolk to sleep. 'Many of the people of fashion are lulled to rest by soft music, or Arabian tales, which their women are taught to repeat.'[63] Russell found two volumes of this scarce book in Aleppo, with only 280 'Nights', and with difficulty he had a copy made for himself. The first publication in English, translated from Galland's French translation of 1704,[64] was in 1712.

More significant was the Russells' use of Arabic sources, especially medical texts. Volume II of the 1794 edition includes a chapter on Arabic sources for medicine and literature. In particular they cited Abu al-Faraj, also known as al-Isfahani, a man of letters, a political and social historian, musicologist, genealogist, philologist and poet who grew up in Addasid Baghdad. Other Arabic sources include Ibn Sina (d. AD 1037), the most renowned mediaeval Islamic philosopher, also famous as a physician, and known in the West as Avicenna;[65] Al Makin's *History* (AD 1123); Ibn Shaddad (AD 1145-1234), a religious scholar and *qadi* who was born in Mosul and died in Aleppo. Ibn Shaddad was closely associated with Saladin (d. AD 1193), and wrote Saladin's *Life* (finished in AD 1228). Patrick also cites other Arabic sources from the Escorial Library in Spain, including *Abulfeda descriptio Aegypti, arabice et latine* (1273-1331)[66] and Ibn al-'Adim (AD 1192-1262), who came from a distinguished Aleppine family and wrote a history of Aleppo, entitled *Zubdat al-Halab*, and a biographical dictionary of notables who had connections with Aleppo entitled *Bughyat al-talab fi ta'rikh Halab*.

The Russells' use of Arabic in footnotes shows that they were familiar with the language. This directly contradicts Edward Lane's statement that 'the author was not sufficiently acquainted with the Arabic language'.[67] Some Aleppines had valuable collections of texts and the Russells, with the help of their friend Tarblos Effendee, Mufti of Aleppo, collected and exported many valuable books from Aleppo.[68] They may well have had access to the Ahmadiyya College which was founded with 3,000 volumes.[69] Indeed, the Russells were part of a line of European Oriental scholars who had associations with Aleppo, including Edward Pococke (1604-91), the greatest Arabist of the 17th century who lived in Aleppo from 1630 to 1636; Bishop Frampton, a chaplain in Aleppo who spent his time learning Arabic and socialising with Aleppines and who remained in the city during the plague of 1667-70; and from 1671-81 the Aleppo chaplain Robert Huntingdon (1637-1701), who also formed a collection of Arabic manuscripts, now in Merton College, University of Oxford.

## Mercantile Gentlemen

> The former [mercantile Gentlemen] often possessed the advantage of speaking Arabic, but were little versed in Natural History and Antiquities.[70]

Both Alexander and Patrick worked in association with

TRAVELLERS IN THE NEAR EAST

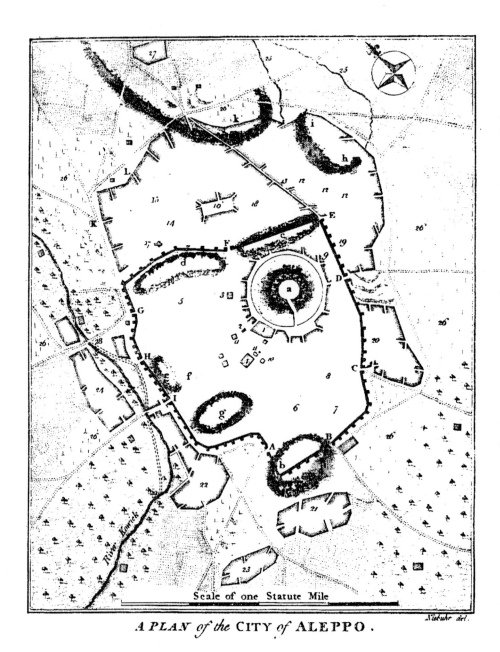

**Figure 3:** A plan of the City of Aleppo *drawn by Carsten Niebuhr and printed opposite page 13 in* The Natural History of Aleppo *(1794), Vol. I.*

# MERCANTILE GENTLEMEN AND INQUISITIVE TRAVELLERS

**Figure 4:** *Aleppo merchants and officials.* Plate II *in* The Natural History of Aleppo *(1794), Vol. I, opposite p. 102.*

the Levant Company. Their brother, William, was its Secretary and a Fellow of the Royal Society of London. William may well have been instrumental in securing their posts; he was certainly one of those who nominated Alexander Russell to the Royal Society of London in 1756.

The Levant Company (The Governor & Company of Merchants of England trading into the Levant Seas), or Turkey Company as it was known by the 18th century, was a firm of English factors who had been operating in the Ottoman Empire from 1581,[71] though permission to trade in the empire had been obtained by the English as early as 1553. Its main factory was in Constantinople, with depots in Smyrna (modern Izmir) and Aleppo, the last considered a haven for British merchants. As 'Ali Bey later recounted, Aleppo 'being continually frequented by a crowd of Europeans and strangers from every nation on account of its commerce, it is almost as well known as any European city.'[72]

There was a strong if isolated community of European merchants residing in the *khans* of Aleppo. It was described in detail by Alexander Russell. The Levant Company merchants led a leisurely life. Russell details their feasting, drinking and hunting: 'The Europeans' table is well stocked with a marvellous assortment of game and fish for which guests were prepared by draughts of weak punch before dinner.'[73] Alexander Russell recounts that 'the Merchants commonly dine in their apartments in the Khanes; some have victuals sent from their own kitchen, but many content themselves with bread, cheese, and fruit, or perhaps a Kabab from the bazar. Their chief repast is supper, at their own houses.'[74] He describes local foods in detail with a list of 141 local dishes.[75]

Alexander Russell would have known many of the Levant Company officials in Aleppo, including Arthur Pollard, consul in 1750; Arthur Radcliffe, in Aleppo from 1734-43 with Radcliffe & Stratton, a man who was stolid and amiable with his family, but severe out of it.[76] He would have known Richard Stratton (d. 1759) who lived in Aleppo for sixteen years, first with the Radcliffe factorage in Aleppo, then as part of Stratton & Hammond from 1749. From his correspondence in the Levant Company records in the Public Record Office (London) Stratton appears competent, excessively self-confident, with a genuine concern for the London merchants he represented. On the other hand he was 'irascible, quick to take offence and quick to forgive, sharp-minded, growing from uncertainty to over confidence in his own abilities'.[77] In the 1740s Aleppo merchant houses included Frye & Mitford, the ailing H.J. & T. March, the great trader William Bellamy,[78] and William Hammond, a member of an old family of Levant merchants who came to Aleppo in 1747 and stayed until 1754, and who was always content with his own judgement He was succeeded in the factoring business by Colvill Bridger, more volatile than Hammond and the last of Radcliffe's factors in Aleppo before the business closed. The Russells would have met a large influx of new traders in 1753-4.

Bridger, a weak man troubled by conflicting interests, was in Aleppo in 1765 (and possibly died shortly after) when he was the Levant Company's Treasurer there.[79] Patrick Russell would also have known John Radcliffe, a cheerful soul fresh from Eton who set out for Aleppo in March 1758 to partner Colvill Bridger. He became heir to the Radcliffe business in London in 1760 and returned home after the death of his elder brother. A spendthrift, he died insolvent and childless in 1783. Others were Dudley Foley with Charles Lisle; Samuel Medley and Arthur Pullinger whose firm went bankrupt; Toby Channer who was apprenticed to William Hayter; Thomas Lansdown, factor in around 1755; and Samuel Bosanquet, a third generation Levant merchant in 1765. Russell would also have known the wealthy Shukri 'A'ida, Christian interpreter at the English Consulate in the 1750s-60s.[80]

The 1730s and 1740s, when Alexander Russell was working in Aleppo, were difficult years to make a large fortune there.[81] English trade was declining by the 1760s for in 1738-47 there were 19 companies of factors and in 1748-57 there were 58 active traders. Most of those working with the Levant Company stayed seven to ten years in the East but they had a conservative social circle, few learnt Arabic or Turkish, and few associated outside their own community.[82] Many continued to make profits, partly through their associations with wealthy and respected merchant families, and returned to London in their early thirties, having accumulated a small fortune via commissions and miscellaneous gains on transactions.[83] Levantine merchants moved in the highest circles in the City of London, even marrying the daughters of gentry and peers.

Bartholomew Plaisted was certainly familiar with Alexander Russell's *Aleppo*. Plaisted was an engineer and surveyor with a quick temper who was dismissed from the East India Company in 1748 and returned to England from India. He crossed the desert to Aleppo in the company of Mr Falquir, a Frenchman he met in Busserah, with whom he had to communicate in Portuguese. He estimated that the caravan he travelled with from Basra to Aleppo included 2,000 camels. They were later joined by the Baghdad caravan, which totalled 5,000 camels (about 400 laden, the rest for trade in Aleppo) and 1,000 men.[84] He reached Aleppo on 23 July 1750 and left Scanderoon (Alexandretta) on 11 August 1750.[85] His short description is a summary extracted from the 1757 version of *Aleppo*: he 'had the pleasure of being acquainted' with Alexander Russell.[86] His summary is even entitled 'A description of Aleppo, and the adjacent country',[87] the same title as Alexander's. Plaisted appears to tire of the task of summarising the third chapter of *Aleppo* (1757) which gives considerable detail about the range of animals, crops and fruit available; from fat-tailed sheep and goats, to hares, gazelles, crops and vegetables: 'With respect to the vegetables of this country, Turky [sic] wheat, barley, cotton, lentils, beans, cicers, Turky [sic] millet, a green kidney bean, musk-melons, water-melons, a small cucumber, bastard saffron, hemp, and several others, they sow in the fields; but about Aleppo they sow no oats, their horses being all fed with barley.'[88] Plaisted noted that 'It would take up too much room to describe the vast variety of fine flowers, herbs and plants to be met in these parts, and a catalogue of their names only would be very tedious, for which reason we shall omit them.'[89] Russell had given an excellent account anyway.

John Carmichael (late Gunner at Anjengo), dismissed from service in the East India Company at Bombay as a troublesome character, was obliged to take the desert route via Syria from London to India in order to settle his affairs. Subsequently he worked for various local Indian rulers and died at Surat. Setting off from Aleppo, he made a credible survey of the 520 miles to Basra. His caravan included 50 horses, 30 mules and 1,200 camels, 600 of them carrying goods worth £300,000. He was a friend of Dr Russell, and when encamped near the village of Irzi on 6 November 1751, reported: 'Had my friend [Alexander Russell] seen him' ('a great fat fellow, a Sheikh', who came to Carmichael's tent in search of a doctor), 'I am persuaded he would admit my knowledge in therapeuticks.' In the absence of Russell, Carmichael bled him with a rusty razor. 'After the operation he slept for about three hours, got up, broke wind, ate a large dish of pillaw, and found himself perfectly recovered.'[90]

Major James Rennell[91] was also indebted to Patrick Russell for 'The Journal of Mr. Carmichael's route across the Great Desert between Aleppo and Basrah, in 1751'[92]. In a paper published in the *Philosophical Transactions of the Royal Society of London*, read to the Royal Society on 17 March 1791, Rennell acknowledged 'the manuscript copy of his [Carmichael's] journal was obligingly communicated by my friend Dr Patrick Russell.'

*Inquisitive Travellers*

> [Inquisitive travellers] though better qualified for inquiry by preparatory studies [than mercantile gentlemen], may be supposed from ignorance of the language, to have sometimes led into error by the menial servants, on whose fidelity, as Interpreters, they are usually obliged to rely.[93]

The earliest mediaeval European account used by Patrick Russell is that of an Iberian Jew, Rabbi Benjamin ben Jonah of Tudela who described Aleppo between 1160 and 1173. He was the first European traveller to approach the frontiers of China[94] – though his translator, Baratier, thought that Benjamin never went to Aleppo. Other travellers cited include Antonio Tenreiro, who was one of the first Portuguese to travel from Aleppo to Basra, in 1523 and back in 1528; and Pierre Belon du Mans (c.1517-65),[95] French naturalist and traveller, who made a tour of the eastern Mediterranean including Aleppo from 1546-9, in order to identify animals, plants and places and objects given in classical writings such as Aristotle.[96] He describes the plants, animals and geographic features of the region.

Venetian and Genoese merchants traded in Aleppo before the 17th century, and while Venice continued to trade with the Orient, the city remained the centre of European civilization. Patrick Russell cites Frederic Caesar, the Venetian who visited Aleppo around 1563 *en route* to India, and after travelling many years returned in 1581 and provided much detail on manners and customs. Other early European travellers they read included Rauwolff (1573);[97] Prosper Alpinus (1553-1616), who provided valuable additions to medical history after his travels in the region between 1581 and 1584, and who was also interested in its natural history;[98] and John Eldred, who resided in Aleppo in the 1580s, and travelled with Ralph Fitch and John Newberrie, setting off in 1583 via the Euphrates for East India accompanied by the jeweller William Leedes, and a painter, James Story.[99] The Russells' description of Arab women and dress is similar to that of Eldred. Whilst exploring the introduction of coffee to the Levant,[100] Patrick studied the intelligent merchant John Sanderson's account of his residency in the East 1584-1602 when he lived in Constantinople between 1592 and 1598 and travelled to Syria. He also searched for details of the early use of tobacco and coffee in *The Preacher's Travels through Syria, Persia etc* (London, 1611), written by an observant Revd John Cartwright who accompanied a London merchant, John Mildenhall, from Aleppo to Persia in 1600. Patrick drew upon the travels of the poet George Sandys (1578-1644), who made a visit to the city in 1611.[101] Another of their major English sources, Dr Richard Pococke (1704-65), later Bishop of Meath, was a wealthy clergyman in Aleppo in 1743 who painted an idyllic picture of English life in Aleppo, and wrote *A Description of the East* in 1763.[102] Threatened by Bedouin raiders when travelling in a caravan he wrote: 'I treated them with coffee, made them my friends and refused to pay anything.'

The strong links between Scotland and the Continent meant that many educated Scots were familiar with texts in European languages. The Russells used many other European travellers' accounts of Aleppo, including Jean de Thévenot (1633-67), a Frenchman who journeyed from Turkey to Egypt in 1656, made a subsequent trip to Persia via Saida, Damascus, Mosul and Baghdad in 1663 and published *Voyage au Levant* in 1664. After his death in Persia in 1667, the *Voyages de M. Thévenot*, 5 vols, was published in 1689.[103] The *Mémoires* of Le Chevalier d'Arvieux, who was in Syria in 1664 and was French consul at Aleppo from 1679 to 1686,[104] contain curious information and accurate observations of the character and political conduct of Turks; Patrick considered his description of Aleppo in the sixth volume to be impartial, and regularly cited d'Arvieux as an authority on a variety of topics, for 'his authority I consider as very respectable'.[105] Patrick cited the French savant, comte Contantin Francois Chassboeuf Volney (1757-1820), who visited the city in 1783, describing its people, politics and antiquities in *Voyage en Égypte et en Syrie* in 1787.

How many other travellers passing through Aleppo during the mid-18th century met the Russells we may never know, though we can identify several of them. Carsten Niebuhr[106] (1733-1815) a German geographer and traveller, and sole survivor of the first scientific expedition to Arabia sponsored by King Frederick V of Denmark, was a colleague. Niebuhr used Carmichael's account of Aleppo, as Carmichael had, later, used the Russells' text. Niebuhr is cited by Patrick on various matters from fashion to dance, and Niebuhr even provided a plan of the city of Aleppo opposite page 13 in the 1794 edition of *Aleppo* (Figure 3): 'In this plan, which I received from my esteemed friend Mr Nieburh [sic], with permission to make whatever use of it I thought fit.'[107]

One could continue to cite many other friendships, such as that between Patrick Russell and James Bruce (who travelled in the Middle East and North Africa between 1768 to 1773) that reinforce the scholarship of the Russell brothers and shed further light on the interaction of European travellers to Aleppo. Yet Patrick also recognised the drawbacks of relying on reports of travellers: 'while from the mode of travelling, and their short stay in places, such matters were left unexplored, as, requiring a greater length of time to investigate, more naturally became fit objects for persons resident in the country.'[108]

*After* The Natural History of Aleppo

In 1760 Alexander Russell obtained his Licence of the Royal College of Physicians of London and his MD degree from the University of Glasgow and in 1760 was appointed a physician to St Thomas's Hospital, London, at that time located near the present site of Fenchurch Street railway station,[109] where he was a colleague of the haughty physician and poet Dr Mark Akenside. In November 1768, Alexander died suddenly of 'putrid fever' at his house at No. 1 Church Court, Wallbrook, despite the efforts of his old friends, John Fothergill and William Pitcairn. Fothergill published a tribute to 'our Russell' presented at a meeting of the Society of (Licentiate) Physicians in October 1769. a society in which Alexander Russell had been active.[110]

> Fothergill dwells upon his even, cool and consistent temper, polite without flattery, with a freedom of behaviour as remote from confidence as from constraint, disinterested and generous. His mind was imbued with a just reverence for God and with duty towards his fellows; a gentleman, without reproach.[111]

Patrick Russell eventually left Aleppo and settled in London in 1772. Alexander Russell obtained additional qualifiication including a Licence from the Royal Society of London in 1777. Even before he left Aleppo Patrick had shown an interest in marine life and had published an account of a 'remarkable marine production' in 1761.[112] In 1781 he went to Vizagapatam in India, accompanying a younger brother, Claud. There Patrick served the East India Company as a botanist and physician in the Carnatic from 1785. In the Carnatic he made large collections of specimens and drawings of plants, fish and reptiles, which on his retirement in 1789, were left to a museum in Madras. After Patrick retired from the East India Company he proceeded to publish many of his findings. He wrote a preface to William Roxburgh's *Plants of the Coronadel Coast* (1795) and went on to publish *An Account of Indian Serpents* (1796).[113] This was followed by *A Continuation of An account of Indian Serpents...* issued in parts between 1801 and 1809,[114] and *Descriptions and Figures of Two Hundred Fishes* (1803).[115] He died on 2 July 1805, having never married. On his legacy, Patrick wrote in his Preface: 'How far the Author's abilities have been equal to the task he has undertaken, the Public will judge; and he intreats their candour.'[116]

*Conclusion*

Before 1801, the core of European scholarship was based on classical traditions. As more classical and mediaeval Arabic sources were translated into European languages and as more people travelled and wrote about their experiences, so Middle Eastern studies in the West developed. In *Orientalism* Edward Said claimed that it was through this study of classical texts that Europe articulated its vision of the Orient, rather than through any 'actual encounter with the real Orient'.[117] He also noted that the early Orientalists, often lawyers or doctors, understood that the route to a proper knowledge of the Orient began with a thorough study of classical texts.[118] Their Orient helped 'to define Europe', not as a 'contrasting image'[119] but rather as a repository of lost or useful knowledge. The incorporation of classical texts and their descriptions of the ways of life meant that the Orient was to become 'an integral part of European material civilisation and culture'.[120]

This essay has explored the often-ignored, yet considerable scholarly contacts between East and West in a 'Pre-Orientalist' era in Aleppo, reflecting on the continuing transmission of scientific expertise up to the end of the 18th century. Like their academic predecessors, Alexander and Patrick Russell played their small yet significant part in reintroducing mediaeval texts, available only in Arabic, back to Europe, along with various Arabic classical manuscripts. For the Russells this scholarly approach also embraced a study of other Oriental cultures and languages, as well as the study of earlier travellers' reports. From the late 16th to the end of the 18th century a few European travellers visited Aleppo, leaving as many as 23 published accounts of the city. The Russells relied not only on their own long-term experiences but on classical and Arabic texts, and earlier travel writing. This begs the question whether their written descriptions hold more value and efficacy than many other travel reports of the area. By using such a wide variety of sources from Syria and Europe, as well as intimate details of the city in which they had worked for years, do the Russells' effectively identify for us the 'real' Orient, if this is possible?

REFERENCES

1. *The Natural History of Aleppo: containing a description of the city, and the principal natural productions in its neighbourhood; together with an account of the climate, inhabitants, and diseases; particularly of the plague, with the methods used by the Europeans for their preservation* (London: printed for the bookseller, Andrew Millar, 1756), Hereinafter *Aleppo* (1756).
2. Edward Said, *Orientalism* (London: Routledge & Kegan Paul, 1978), pp. 159-60.
3. Sari J. Nasir, *The Arabs and the English* (London: Longman, 1979), p. 63.

4. It was produced in London and printed for G.G. and J. Robinson (bookseller) in 2 volumes in 1794, with magnificent plates and a map (26 x 32 cm). This was reissued in 1856 in one volume. Reprinted versions are available published by Aldershot, Gregg International Publishers, 1969, 2 vols. Hereinafter *Aleppo* (1794).
5. Sarah Searight, *The British in the Middle East* (London: East-West Publications, 1979), p. 231; B.C. Corner and C.C. Booth (eds), *Chain of Friendship: selected letters of Dr John Fothergill of London, 1735-1780* (Cambridge, Mass: Harvard University Press, 1971), 290n.
6. *Aleppo* (1794), I, p. v.
7. S.M. Devlin-Thorp (ed.) *Fellows of the Royal Society of Edinburgh Vol 2: Literary Fellows Elected 1783-1812* see http://www.amazon.co.uk/exec/obidos/ASIN/090769201X/qid%3D1024674075/202-3022236-6247062.
8. Cuming later set up in practice in Dorchester in 1739. Fothergill proposed that Cuming move to London after Russell died in 1768 but Cuming preferred to remain in Dorchester.
9. Corner and Booth (1971).
10. George Cleghorn, *Observations on the Epidemical Diseases in Minorca. From the year 1744 to 1749. To which is prefixed, a short account of the climate, productions, inhabitants, and endemial distempers of that island* (London: D. Wilson, 1751).
11. Letter to Lieutenant-Colonel Gilbert Ironside 22 December 1774, in Corner and Booth (1971), p. 433. In 1751 Cleghorn settled in Dublin, by 1771 he had become a lecturer in anatomy in Trinity College, Dublin (ibid, p.343), was made Professor of Anatomy in 1785 and was an original member of the Royal Irish Academy. Hingston Fox, *Dr John Fothergill and his Friends: chapters in eighteenth-century life* (London: Macmillan, 1919), pp. 121-7.
12. He wrote papers on palsy, hydatids, general emphysema, the use of corrosive sublimate and of mezereon in syphilis, Fox (1919), p. 120.
13. Wadi' 'Abd Allah Qastan as *al-Ifranj f Halab* [Foreigners in Aleppo], published, romanised in 1969.
14. *A Compendium of the most Approved Modern Travels: containing a distinct account of the religion, government, commerce, manners, and natural history, of several nations, illustrated and adorned with many*

*useful and elegant copper-plates* (London: published for John Scott, 1757). Russell's section hereinafter as *Aleppo* (1757). Also translated into French by Philippe-Florent de Puisieux (1713-1772), *Les Voyageurs modernes, ou, abrégé de plusieurs voyages faits en Europe, Asie & Afrique, traduit de l'Anglois* (Paris: chez Nyon, Guillyn & Hardy, 1760), in vol. IV as 'Voyage d'Alexandre Drummond ... en Chypre & en Syrie. Description d'Alep & des pays voisins'.

15. Alexander Drummond, *Travels through different cities of Germany, Italy, Greece, and several parts of Asia as far as the banks of the Euphrates* (London: printed by W. Strahan for the author, 1754).

16. *Aleppo* (1794), I, p. xix.

17. 4 volumes. Volume 1. 'A journey from Aleppo to Jerusalem', by H. Maundrel [Henry Maundrell]; The travels of Dr. Thomas Shaw 'An account of a journey to Palmyra, otherwise Tedmor in the desart'. Volume II. The travels of Dr. Richard Pococke; The travels of Alexander Drummond, Esq.; 'A description of Aleppo and the adjacent country', by A. Russel [Alexander Russell]; 'The travels of Mr.[Jonas] Hanway through Russia, Persia, and other parts of Europe, for settling a trade upon the Caspian Sea' (chap. I-III). Volume III. 'The travels of Mr. Hanway' (chap. IV-XVII); 'The natural history of Norway', by ... E. [Erich] Pontoppidan. Volume IV. 'The Travels of Frederick Lewis [Frederik Ludvig] Norden through Egypt and Nubia'. Also excerpts by Robert Wood. Copies of this 1757 version are in libraries in Leeds, Oxford. A copy Russell's extract, published in London, is in the Library of Congress dated 1762-90.

18. *The World Displayed: or, a Curious Collection of Voyages and Travels, selected from the Writers of all Nations. In which the conjectures and interpolations of several vain editors and translators are expunged; every relation is made concise and plain, and the Divisions of Countries and Kingdoms are clearly and distinctly noted. Embellished with Cuts* (Dublin: James Williams, 1779, 6th edn, corrected), XIII, p. 63-103. The compendium was compiled by Christopher Smart, Oliver Goldsmith and Samuel Johnson and with an introduction by Samuel Johnson.

19. See Searight (1979), p. 89, for copy of 'Plate of Goat and fat-tailed sheep'.

20. Nasir (1979), p. 46.
21. Searight (1979), p. 77.
22. *Aleppo* (1757), pp. 102-103.
23. *Aleppo* (1757), p. 102.
24. H.C. Cameron, *Sir Joseph Banks, the Autocrat of the Philosophers* (London: Batchworth Press, 1952).
25. *Aleppo* (1794), I, p. ix.
26. *Philosophical Transactions of Royal Society*, 57 (December 1767), 117. James Gordon, one of London's well-known nursery gardeners, specialised in the sale of North American plants (Corner and Booth, 1971, p. 305. It flowered for the first time in Fothergill's garden in May 1766 and grew to twelve feet. E.D. Ehret, the painter who illustrated *Aleppo* (1794), gave a paper about it at the Royal Society in 1767, ibid., p. 57, 114.
27. William Pitcairn, a Scot from Fifeshire, studied medicine with Boerhaave in Leyden and in Rheims. From 1750 to 1780 he was a physician at St Bartholomew's Hospital, London. Pitcairn had a fine botanical garden in Upper Street, Islington (William Munk, *The Roll of the Royal College of Physicians of London &c*, London: published by the College, 1878, II, pp. 172-4, quoted in Fox 1919, p. 185 at a time when Fothergill was developing his Upton Park gardens, and collected plants in West Africa with Sir Joseph Banks, the Earl of Tankerville and William Brass.
28. For example, the *Sirat Salah ad-Din* [Life of Saladin], finished by 1228 by *qadi* Ibn Shaddad (AD 1145-1234), using a Latin translation of 1732.
29. Charles Perry, *A View of the Levant: particularly of Constantinople, Syria, Egypt, and Greece* (London: T. Woodward, 1743; repr 1763), p. 141.
30. There are several excellent modern sources on 18thcentury Aleppo: *The Desert Route to India* (London: Hakluyt, 1928; repr. Nendeln: Kraus Reprint, 1967), edited by Douglas Carruthers; Ralph Davis, *Aleppo and Devonshire Square* (London: Macmillan, 1967); Abraham Marcus, *The Middle East on the Eve of Modernity: Aleppo in the eighteenth century* (New York: Columbia University Press, 1989); Bruce Masters, *The Origins of Western Economic Dominance in the Middle East: mercantilism and the Islamic economy in Aleppo, 1600–1750* (London:

London University Press, 1988) and Eldem Edhem, Daniel Goffman, and Bruce Masters, *The Ottoman City between East and West* (Cambridge: Cambridge University Press, 1999) amongst others.

31. A.C. Wood, *A History of the Levant Company* (1935; repr. London: Cass, 1964). The route was superseded when goods were transported by the East India Company via the Cape of Good Hope instead of the overland route.
32. *Aleppo* (1794), I, p. 199.
33. Russell describes Aleppine textile workshops in *Aleppo* (1794), I, p. 161.
34. *Aleppo* (1757), p. 71.
35. *Aleppo* (1757), p. 71.
36. *Aleppo* (1794), I, p. 100-15.
37. *Aleppo* (1794), I, 235; Marcus (1989), p. 31.
38. Russell, Patrick, *A Treatise of the Plague*, 1791, p. 9.
39. Marcus (1989), p. 356.
40. *Aleppo* (1794), I, pp. 338-9; also mentioned by Drummond (1754), 182, Contantin-François Chassboeuf, comte de Volney, *Voyage en Syrie et en Égypte pendant les années 1783, 1784 et 1785*, 2 vols (Paris: Volland et Desenne, 1787), II, p. 135.
41. Fox (1919), p. 120.
42. *A Treatise of the Plague: containing an historical journal, and medical account, of the plague, at Aleppo, in the years 1760, 1761 and 1762. Also, remarks on quarantines, lazarettos, and the administration of police in times of pestilence. To which is added, an appendix, containing cases of the plague; and an account of the weather, during the pestilential season*, a Government publication, printed in London for the booksellers, G.G.J. and J. Robinson, 1791, 12 p. l., p. 583, clix p. 4l ; 30 x 25 cm, see Marcus (1989), p. 51.
43. *Treatise* (1791), pp. 1-70.
44. Especially *Treatise* (1791), pp. 1-70, 280, 337.
45. *Aleppo* (1794), II, pp. 336-8.
46. *Aleppo* (1794), I, pp. 298-333, 344-5.
47. *Aleppo* (1794), II, pp. 79, 83.
48. *Philosophical Transactions of the Royal Society of London*, 58 (December 1768), pp. 142.

49. 'An Account of the Late Earthquakes in Syria', *Philosophical Transactions of the Royal Society of London*, 51 (December 1760), pp. 529-34.
50. Davis (1967), p. 75.
51. *Aleppo* (1794), I, p. xiii.
52. Marcus (1989), p. 379.
53. *Aleppo* (1794), I, p. ix.
54. Sale catalogue of Fothergill's library is in the Library of The Society of Friends in London. Corner and Booth (1971), p.21.
55. *Aleppo* (1794), I, p. xiii
56. ibid., I, p. xiii.
57. ibid., I, p. 383.
58. ibid., I, 368.
59. ibid., I, 373, 367.
61. ibid., I, 383.
61. ibid., I, 368, 378.
62. ibid., 148-50.
63. ibid., 81.
64. Antoine Galland, *Les Mille et une nuit: contes arabes*, 8 vols (Le Haye: chez Pierre Husson, 1704-1728).
65. *Aleppo* (1794), I, pp. 372-3, for example.
66. Edited by Joannes David Michaelis (Gottingen: Dieterich, 1776).
67. Edward W. Lane, 'Preface', *An Account of the Manners and Customs of the Modern Egyptians* (London: 1836; repr. London: Ward, Lock, 1890).
68. *Aleppo* (1794), I, p. 337.
69. Marcus (1989), pp. 237-8.
70. *Aleppo* (1794), I, p. ix.
71. Wood (1964), esp. Ch. 11.
72. *Travels of Ali Bey* (London: Longman, Hurst, Rees, Orme, and Brown, 1816), Vol II, written 1803-07.
73. Searight (1979), p. 35 quoting *Aleppo* (1794).
74. *Aleppo* (1794), I, pp. 143-4.
75. ibid., I, pp. 73-94, 115-19, 172-7.
76. Davis (1967), p. 19.
77. Searight (1979), pp. 19, 25, 67.

78. Davis (1967), p. 54.
79. ibid., 21, p. 2.
80. Marcus (1989), p. 366.
81. Davis (1967), p. 19.
82. ibid., p. 79.
83. ibid., p. 20.
84. Carruthers (1928), p. xxxiii.
85. Bartholomew Plaisted, 'Narrative of a Journey from Basra to Aleppo in 1750' in Carruthers (1928), pp. 59-128. It was translated into French in 1758 and this was reprinted in Thomas Howel, *Voyage en retour de l'Inde par une route en partie inconnue jusqu'ici* (Paris: de l'Imprimerie de la République, an V 1796-1797).
86. Preface, see Carruthers (1928), p. 58.
87. Plaisted (1757), pp. 107-13.
88. *Aleppo* (1757), p. 101.
89. Carruthers (1928), p. 113.
90. John Carmichael, 'Narrative of a Journey from Aleppo to Basra in 1751', in *The Desert Route to India* (1772; London: Hakluyt, 1928), p. 149.
91. *Treatise on the Comparative Geography of Western Asia* (London, 1831). Altas.
92. John Carmichael, of the East India Company, *A Journal from Aleppo, over the desart to Basserah, October 21, 1771* (1751), in British Library.
93. *Aleppo* (1757), p. 101.
94. Benjamin ben Jonah of Tudela, *Voyages de Rabbi Benjamin fils de Iona de Tudele... depuis l'Espagne jusqu'a la Chine: Ou l'on trouve plusieurs choses remarquables concernant l'histoire & la geographie & particulierement l'etat des Juifs au douzieme siecle. Traduits de l'Hebreu et enrichis de notes et de dissertations historiques et critiques sur ces voyages par J.P. Baratier* (Amsterdam, 1734).
95. *Les Observations de plusieurs singularitez & choses memorables: trouuées en Grece, Asie, Indée, Egypte, Arabie, & autres pays estranges, redigées en trois liures* (Paris: chez Guillaume Cauellet, 1555) [Observations of Several Curiosities and Memorable Objects...].
96. Serge Sauneron, ed., *Le Voyage en Égype de Pierre Belon du Mans*

*1547* (Cairo: Institut français d'archéologie orientale (IFAO), 1970).
97. John Ray, *Collection of Curious Travels, &c.* (London: printed for S. Smith and B. Walford, 1705). Patrick Russell used vol. II of the 1738 edition in *Aleppo* (1794), I.
98. *De Plantis Aegypti Liber Accessit etiam liber de Balsamo*, 2 vols (Venice: apud Franciscum de Franciscis Senensem, 1592), [Book of Egyptian Plants] and *De Medicina Aegyptiorum Libri IV* (Venice: apud Franciscum de Franciscis Senensem, 1591). See also Prosper Alpini, *La médecine des egyptiens;* traduite du latin, présentée et annotée par R. de Fenoyl; index de Marcelle Desdames (Cairo: IFAO, 1980).
99. Richard Hakluyt, *The Principal Navigations, Voyages, etc*, 5 vols (1599-1600; repr. Glasgow: J. MacLehose 1903-1907).
100. Coffee was exported to Europe from Aleppo from the 1650s but this supply was replaced in around 1730 with West Indian coffee which was even exported to Egypt. Egypt had previously obtained all supplies from Mocha and exchanged coffee for cloth in the markets of Cairo with the English and French.
101. George Sandys, *Relation of a Journey begun Ann. Dom: 1610: Fovre bookes, containing a description of the Turkish Empire, of Aegypt of the Holy Land, of the remote parts of Italy, and ilands adioyning* (London: printed for W. Barrett, 1615; repr. Amsterdam: Theatrum Orbis Terrarum; New York: Da Capo Press, 1973).
102. *A Description of the East and some other countries*, 5 bks, 2 vols in 3 (London: printed for author, 1743-5), II, part 1, pp. 150-2; MSS 22995, 22997-8 in British Library. Travelling in region 1737-8.
103. Jean de Thévenot, *The Travels of Monsieur de Thévenot into the Levant. In three ... parts: I. Turkey. II. Persia. III. The East-Indies*, transl. Archibald Lovell (London: printed by H. Clark, for H. Faithorne, J. Adamson, C. Skegnes, and T. Newborough, 1687), II, pp. 31-32.
104. Le Chevalier Laurent d'Arvieux, *Mémoires du Chevalier d'Arvieux*, 6 vols, ed. Jean-Baptiste Labat (Paris: chez Charles-Jean-Baptiste Delespine le fils, 1735), VI, pp. 411-28. See also Régine Goutalier, *Le Chevalier d'Arvieux: Laurent le Magnifique: un humaniste de belle humeur* (Paris: L'Harmattan, 1997).
105. *Aleppo* (1794), I, p. 351.

106. Carsten Niebuhr, *Voyage en Arabie et en d'autres pays circonvoisins; traduit de l'allemand*, 3 vols, maps (Amsterdam: S.J. Baalde, 1780).
107. *Aleppo* (1794), I, p. 12.
108. ibid, I, p. ix.
109. James Gordon, to whom Alexander Russell sent seeds from Aleppo, also had a seed shop in Fenchurch Street. Gordon also supplied Fothergill; see Corner and Booth (1971), p.302; Fox (1919), p. 193.
110. John Fothergill, 'Essay on the Character etc' (1769) in J. Fothergill, *The Works of John Fothergill*, ed. J.C. Lettsom, 3 vols (London: printed for C. Dilly in the Poultry, 1783-4), II, pp. 361-79.
111. Fox (1919), p. 120. There is also a stippled portrait of Alexander Russell by Trotter in J.C. Lettsom, *Memoirs of John Fothergill, M.D. &c*, 5 vols (London: printed for C. Dilly, 1786).
112. *Philosophical Transactions of the Royal Society of London*, 52 (December 1761), p. 554.
113. *An account of Indian serpents, collected on the coast of Coromandel; containing descriptions and drawings of each species; together with experiments and remarks on their several poisons*, Presented to the Hon. The Court of Directors of the East India Company, and published by their order, under the superintendence of the author (London: Printed by W. Bulmer and Co. for George Nicol, 1796).
114. Including *A Continuation of An account of Indian Serpents: containing descriptions and figures from specimens and drawings, transmitted from various parts of India to the East India Company* (London: Printed by W. Bulmer and Co., for G. and W. Nicol, 1801) for the East India Company.
115. *Descriptions and Figures of Two Hundred Fishes: collected at Vizagapatam on the coast of Coromandel*, 2 vols (London: G. and W. Nicol, 1803).
116. *Aleppo* (1794), I, p. xix.
117. Said 1978, p. 80.
118. ibid., p. 79.
119. ibid., p. 2.
120. ibid., p. 2.

# BIBLIOGRAPHY

Abu al-Fida' Isma`il ibn 'Ali, *Abulfedae Descriptio Aegypti, arabice et latine/Ex codice parisiensi edidit, latine vertit, notas adiecit*, edited by Joannes David Michaelis (Goettingae: apud Joann Christian Dieterich, 1776).

'Ali Bey, *Travels of Ali Bey: in Morocco, Tripoli, Cyprus, Egypt, Arabia, Syria, and Turkey: between the years 1803 and 1807* (London: Longman, Hurst, Rees, Orme, and Brown, 1816).

Alpini, Prosper, *De medicina Aegyptiorum, libri quatuor: In quibus multa cum de vario mittendi sanguinis usu per venas, arterias, cucurbitulas ac scarificationes nostris inusitatas, deque inustionibus, & aliis chyrurgicis operationibus, tum de quamplurimis medicamentis apud Aegyptios frequentioribus, elucescunt: Quae cum priscis medicis doctissimis, olim notissima, ac pervulgatissima essent, nunc ingenti artis medicae iactura à nostris desiderantur* (Venice: apud Franciscum de Franciscis Senensem, 1591).

—, De plantis Aegypti liber: *In quo non pauci, qui circa herbarum materiam irrepserunt, serrores, deprehenduntur, quorum causa hactenus multa medicamenta ad vsum medicine admodum expetenda, plerisque medicorum, non sine artis iactura, occulta, atque obsoleta iacuerunt. Accessit etiam Liber de balsamo aliàs editus* (Venice: apud Franciscum de Franciscis Senensem, 1592).

—, *La médecine des egyptiens;* traduite du latin, présentée et annotée par R. de Fenoyl; index de Marcelle Desdames (Cairo: Institut français d'archéologie orientale 1980).

Belon du Mans, Pierre, *Les Observations de plusieurs singularitez & choses memorables: trouuées en Grece, Asie, Iudée, Egypte, Arabie, & autres pays estranges, redigées en trois liures* (Paris: chez Guillaume Cauellet, 1555).

Benjamin ben Jonah, of Tudela, *Voyages de Rabbi Benjamin fils de Iona de Tudele ... depuis l'Espagne jusqu'a la Chine: Ou l'on trouve plusieurs choses remarquables concernant l'histoire & la geographie & particulierement l'etat des Juifs au douzieme siecle. Traduits de l'Hebreu et enrichis de notes et de dissertations historiques et critiques sur ces voyages par J.P. Baratier* (Amsterdam, 1734).

Cameron, H.C., *Sir Joseph Banks, the Autocrat of the Philosophers* (London: Batchworth Press, 1952).

Carruthers, Douglas (ed.), *The Desert Route to India: being the Journals of four travellers (William Beawes, Gaylard Roberts, Bartholomew Plaisted, John Carmichael) by the Great Desert caravan route between Aleppo and Basra, 1745-1751* (London: Hakluyt, 1928; repr. Nendeln: Kraus Reprint, 1967).

Cleghorn, George, *Observations on the Epidemical Diseases in Minorca. From the year 1744 to 1749. To which is prefixed, a short account of the climate, productions, inhabitants, and endemial distempers of that island* (London: D. Wilson, 1751).

—, *A Compendium of the most Approved Modern Travels: containing a distinct account of the religion, government, commerce, manners, and natural history, of several nations, illustrated and adorned with many useful and elegant copper-plates* (London: published for John Scott, 1757).

Corner, B.C. and C.C. Booth (eds), *Chain of Friendship: selected letters of Dr John Fothergill of London, 1735-1780* (Cambridge, Mass: Harvard University Press, 1971).

d'Arvieux, Le Chevalier Laurent, *Mémoires du Chevalier d'Arvieux*, 6 vols, ed. Jean-Baptiste Labat (Paris: chez Charles-Jean-Baptiste Delespine le fils, 1735).

Davis, Ralph, *Aleppo and Devonshire Square* (London: Macmillan, 1967).

Devlin-Thorp, S.M. (ed.) *Fellows of the Royal Society of Edinburgh Vol 2: Literary Fellows Elected* 1783-1812 http://www.amazon.co.uk/exec/obidos/ASIN/090769201X/qid%3D102 4674075/202-3022236-6247062.

Drummond, Alexander, *Travels through different cities of Germany, Italy, Greece, and several parts of Asia as far as the banks of the Euphrates: in a series of letters. Containing, an account of what is most remarkable in their present state, as well as in their monuments of antiquity* (London: printed by W. Strahan for the author, 1754).

Edhem, Eldem, Daniel Goffman, and Bruce Masters, *The Ottoman City between East and West: Aleppo, Izmir, and Istanbul* (Cambridge: Cambridge University Press, 1999).

Fothergill, John, *The Works of John Fothergill (with Memoirs of W. Cuming, G. Cleghorn, A. Russell, P. Collinson)*, ed. J.C. Lettsom, 3 vols (London: printed for C. Dilly in the Poultry, 1783-4).

Fox, Hingston, *Dr John Fothergill and his Friends: chapters in eighteenth-century life* (London: Macmillan, 1919).

Galland, Antoine (trans.), *Les Mille et une nuit: contes arabes*, 8 vols (La Haye: chez Pierre Husson, 1704-28).

Goutalier, Régine, *Le Chevalier d'Arvieux: Laurent le Magnifique: un humaniste de belle humeur* (Paris: L'Harmattan, 1997).

Hakluyt, Richard, *The Principal Navigations, Voyages, etc*, 5 vols (1599-1600; repr. Glasgow: J. MacLehose, 1903-1907).

Howel, Thomas, *Voyage en retour de l'Inde: par terre, et par une route en partie inconnue jusqu'ici; suivi d'Observations sur le passage dans l'Inde par l'Égypte et le grand désert, par James Capper; traduit de l'anglais par Théophile Mandar* (Paris: de l'Imprimerie de la République, an V, 1796-1797).

Lane, Edward W. *An Account of the Manners and Customs of the Modern Egyptians* (London: 1836; repr. London: Ward, Lock, 1890).

Lettsom, J.C., *Memoirs of John Fothergill, M.D. &c*, 5 vols (London: printed for C. Dilly, 1786).

Marcus, Abraham, *The Middle East on the Eve of Modernity: Aleppo in the eighteenth century* (New York: Columbia University Press, 1989).

Masters, Bruce, *The Origins of Western Economic Dominance in the Middle East: mercantilism and the Islamic economy in Aleppo, 1600–1750* London: London University Press, 1988).

Munk, William *The Roll of the Royal College of Physicians of London &c* (London: published by the College, 1878)

Nasir, Sari J., *The Arabs and the English* (London: Longman, 1979).

Niebuhr, Carsten, *Voyage en Arabie et en d'autres Pays circonvoisins; traduit de l'allemand* (Amsterdam: S.J. Baalde, 1780).

Perry, Charles, *A View of the Levant: particularly of Constantinople, Syria, Egypt, and Greece* (London: T. Woodward, 1743; repr. 1763).

Pococke, Richard, *A Description of the East and some other countries*, 5 bks, 2 vols in 3 (London: printed for author by W. Bowyer; and sold by J. and P. Knapton [et al.], 1743-5).

Puisieux, Philippe-Florent de, 1713-1772, *Les Voyageurs modernes, ou, abrégé de plusieurs voyages faits en Europe, Asie & Afrique, traduit de l'Anglois* (Paris: chez Nyon, Guillyn & Hardy, 1760) includes in vol IV 'Voyage d'Alexandre Drummond... en Chypre & en Syrie. Description d'Alep & des pays voisins'.

Ray, John, *Collection of Curious Travels, &c.* (London: printed for S. Smith and B. Walford, 1705; repr. 1738).

Russell, Alexander, *The Natural History of Aleppo: containing a description of the city, and the principal natural productions in its neighbourhood; together with an account of the climate, inhabitants, and diseases; particularly of the plague, with the methods used by the Europeans for their preservation* (London: Andrew Millar, 1756; second edition, 2 vols, London: printed for G.G. and J. Robinson, 1794; repr. Aldershot: Gregg International Publishers, 1969).

Russell, Patrick, 'An Account of the Late Earthquakes in Syria', *Philosophical Transactions of the Royal Society of London*, 51 (December 1760), pp. 529-34.

—, *A Treatise of the Plague: containing an historical journal, and medical account, of the plague, at Aleppo, in the years 1760, 1761 and 1762. Also, remarks on quarantines, lazarettos, and the administration of police in times of pestilence. To which is added, an appendix, containing cases of the plague; and an account of the weather, during the pestilential season* (London: G.G.J. and J. Robinson, 1791).

—, *An account of Indian serpents, collected on the coast of Coromandel; containing descriptions and drawings of each species; together with experiments and remarks on their several poisons*, (London: Printed by W. Bulmer and Co. for George Nicol, 1796).

—, *A Continuation of An account of Indian Serpents: containing descriptions and figures from specimens and drawings, transmitted from various parts of India to the East India Company* (London: Printed by W. Bulmer and Co., for G. and W. Nicol, 1801) for the East India Company.

Russell, Patrick, *Descriptions and Figures of Two Hundred Fishes: collected at Vizagapatam on the coast of Coromandel*, 2 vols (London: G. and W. Nicol, 1803).

Said, Edward, *Orientalism* (London: Routledge and Kegan Paul, 1978).

Sandys, George, *Relation of a Journey begun Ann. Dom: 1610: Fovre bookes containing a description of the Turkish Empire, of Aegypt of the Holy Land, of the remote parts of Italy, and ilands adioyning* (London: printed for W. Barrett, 1615; repr. Amsterdam: Theatrum Orbis Terrarum; New York: Da Capo Press, 1973).

Sauneron, Serge (ed.), *Le Voyage en Égype de Pierre Belon du Mans 1547* (Cairo: IFAO, 1970).

Searight, Sarah, *The British in the Middle East* (London: East-West Publications, 1979).

Smart, Christopher et al., *The World Displayed: or, a Curious Collection of Voyages and Travels, selected from the Writers of all Nations. In which the conjectures and interpolations of several vain editors and translators are expunged; every relation is made concise and plain, and the Divisions of Countries and Kingdoms are clearly and distinctly noted. Embellished with Cuts*, compiled by Christopher Smart, Oliver Goldsmith and Samuel Johnson; introduction by Samuel Johnson (6th edn, corrected, Dublin: James Williams, 1779).

Thévenot, Jean de, *The Travels of Monsieur de Thévenot into the Levant. In three... parts: I. Turkey. II. Persia. III. The East-Indies*, trans. Archibald Lovell (London: printed by H. Clark, for H. Faithorne, J. Adamson, C. Skegnes, and T. Newborough, 1687).

Volney, Contantin-François Chassboeuf, comte de, *Voyage en Syrie et en Égypte pendant les années 1783, 1784 et 1785*, 2 vols (Paris: Volland et Desenne, 1787).

Wood, A.C., *A History of the Levant Company* (1935; repr. London: Cass, 1964).

# CHAPTER THREE
# Jean-Baptiste Adanson (1732-1804): A French Dragoman in Egypt and the Near East

## Jen Kimpton

The French dragoman, Jean-Baptiste Adanson, lived for 50 years in the Near East and North Africa. Almost his entire residency was prior to Napoleon Bonaparte's 1798 invasion of Egypt, after which Europeans were more numerous. Adanson had to operate within the cultural confines of the Ottoman and local authorities, over which France had decreasing influence. This often made his business on behalf of his European homeland difficult, tense and dangerous. In spite of the difficulties however, Adanson still found the time for intellectual pursuits, the results of which can be seen in his handwritten manuscripts and his drawings of local flora, fauna and antiquities which have survived to this day. This chapter considers one illustrated manuscript in particular – a treatise on ancient Egyptian religion, history and hieroglyphics. The ideas and illustrations in this document exemplify the type of pre-decipherment efforts that formed the context for the emergence of Egyptology as an academic field.

*Origins and Education: Adanson in France*

Jean-Baptiste-Félix-Hubert Adanson was born in 1732 in Paris, where his father, Léger Adanson, served as *écuyer* (squire) to the Archbishop Charles-Gaspard-Guillaume de Vintimille du Luc.[1] Jean-Baptiste was the youngest of four Adanson sons: the two eldest had become soldiers, and the third son, Michel Adanson, had been able to obtain a small canonry with which to support his education.[2] Jean-Baptiste's access to an education was due entirely to the intervention of the Archbishop (whose great-nephew was Jean-Baptiste's godfather), who lobbied aggressively for the youth's admittance to the Chambre des enfants de langues (also referred to as the École des jeunes de langues), the annex of the Collège de Louis-le-Grand in Paris dedicated to the training of students to serve as official translators in the French *échelles*[3] of the Levant and Barbary. Consequently, when Jean-Baptiste Adanson was entered into the École des jeunes de langues on 19 September 1740 at the age of eight, his fate – to spend his adult life away from France – was essentially sealed.

Since Adanson's education at the École des jeunes de langues was the lens through which he would perceive his foreign surroundings, it is worthwhile looking briefly at his schooling.[4] France had maintained diplomatic relations with the Ottoman Empire since the end of the 15th century, at first via temporary missions, and then through the establishment of a permanent ambassador in Constantinople in 1535. However, since the French representatives were not trained in Turkish or Arabic, they had to conduct their affairs through the medium of translators (known as dragomen). At that time the only people who had sufficient knowledge of the languages of both the East and West were the descendants of the Genoese families who had remained in their trading colony, Galata, after the fall of Constantinople to the Ottomans. As

French political and commercial interests in the area increased, two important problems with their dependence on these native dragomen became apparent.[5] Firstly, the dragomen and their families were subjects of the Ottoman Empire, and therefore subject to reprisals from the local authorities, a situation which could impede them from acting aggressively on behalf of France. Secondly, the loyalty of the local dragomen to French interests was often quite dubious, and dragomen were known to use their position for their own commercial advantages. In the light of these unsatisfactory conditions, the French state decided to begin training its own dragomen, and so in 1669 the École des jeunes de langues was created with funding from the king himself (Louis XIV) and from the Chambre de Commerce de Marseille.[6]

Initially the school was intended to comprise two sections, with one at Pera (the foreign quarter of Constantinople) and one at Smyrna, both under the direction of the Capuchin monks who had established bases at both sites. Quickly the idea of the twin school at Smyrna was abandoned, and the full contingent of 18 apprentice-interpreters from Paris and Marseille was educated at Pera only. This first stage of the school did not produce satisfactory results, as was witnessed in 1698 by the French *Chargé d'affaires*, J.-B. Fabre, who wrote: 'Je suis obligé de dire... qu'aucune nation n'est plus mal servie que la nôtre, en drogmans'.[7] One of the problems seemed to have been that, despite the initial intention to admit only students no older than nine or ten, in actual practice the age of these first students ranged between 11 and 20, and even higher. But the more serious problem rested within the Capuchins' curriculum and general approach to the education of the future dragomen. The teachers they employed had little in the way of credentials, and no Arabic was taught at all (the Capuchins claimed that there was no suitable tutor for Arabic in Constantinople and, further, they wished to spare the students mental fatigue). Instead the Capuchins offered small amounts of French, Latin, modern Greek, Italian, and Turkish, but even in those areas the graduates of the school at Pera were discovered to be barely competent. In short, after thirty years, the École des jeunes de langues at Pera came nowhere near solving the problems which had motivated its establishment.

In 1692 the Jesuits applied for permission to set up an institution in France for the religious education of foreign students from the Orient with the idea that these students, once converted to the Catholic faith, should return to their home countries to help the Jesuits introduce it to the native populace. Although the idea was received favourably by the king, eventually the concept became linked to the foundering effort to train dragomen to serve France. Thus in 1700 a new Jesuit school was established at the Collège de Louis-le-Grand in Paris with a dual purpose: Greek, Syrian, Arab and Armenian students would come to Paris to be educated at the expense of the state, and in exchange these students were expected to return to their homes either as missionaries for the Jesuits or as dragomen for France.

The Jesuits seemed to have been better teachers than their Capuchin counterparts at the school in Pera, which still remained in operation for French students. Nevertheless, it must be surmised that the Jesuit educators did not emphasise service to the state as heavily as service to religion, since the few foreign students who did graduate and keep their obligations to French interests did so as missionaries, not dragomen. Generally, the Jesuits failed to transform their charges into westerners, and instead the students tended to take advantage of their costly education in France, often reverting to their former faith and becoming merchants upon returning to their home countries. By 1721 the fact that the school at Louis-le-Grand was not producing any suitable dragomen was public knowledge, and a new training method was sought.

Fortunately for Adanson, who would enter the school in 1740, this latest effort to train dragomen adequately for the field would prove more successful, if still not completely adequate. From 1721 it was decided that both the Jesuit school at Louis-le-Grand and the Capuchin École des jeunes de langues at Constantinople would be maintained, but that they would now operate in tandem. Now, only French students would be admitted (born either in France or in the Levant of French parents),[8] and no student over 15 years of age (ten years being the preferred age) would be allowed to start the programme. The students would begin in Paris with the Jesuits, and would study French, Latin (most textbooks were still written in that language), Turkish and Arabic, and remain there for up to ten years before being sent on to the school at Constantinople where they could refine their specialised language skills. At Louis-le-Grand the curriculum was improved through the hiring of specialists (usually not Jesuits) to teach the Oriental languages.

By taking in students at such a young age, the Jesuit *pères* became intimately involved in their students' development into adulthood, and this familial setting along with the small number of students (ten or fewer) may have helped inspire the loyalty in the future dragomen that previous attempts had failed to achieve. The advantages of a free education and a respectable career were attractive to many parents of younger sons, and competition for one of the few spaces that became available each year was fierce. In the end, admission to the school was attained through a nomination by the king (or the French ambassador at Constantinople in the case of candidates from abroad), so it was critical to have an influential sponsor, as Adanson had in the Archbishop de Vintimille du Luc.

Nevertheless, in spite of the improvements undertaken in 1721, there were still serious flaws in the educational programme. The natural bias of the Jesuits for religious training had been ameliorated but not entirely eradicated. The Oriental language professors taught at most for only an hour and a half per day, and even then they were faced with the entire enrollment of the school in the same classroom, so that the students' ages might range from eight to 18. The Jesuit *pères* considered the mastery of Latin to be of paramount importance; a good student was a student who excelled in Latin and Greek. Success in the Oriental languages was not respected as highly. Practical problems such as the scarcity of competent teachers and the difficulty in acquiring dictionaries and grammars continued to dog the enterprise. Thus, it was still all too easy to complete the programme at Louis-le-Grand in Paris and yet possess only a rudimentary understanding of Turkish, the most important language for a dragoman in the Ottoman Empire. Furthermore, the quality of education at the Capuchin college at Pera, where the students who had completed their studies in Paris went to finish their training, had not improved, and students vied with one another to attain their first posts in the French *échelles* simply to escape the poor conditions there as quickly as possible.

This, then, was the situation that Adanson faced when he became a *jeune de langues* at Louis-le-Grande in Paris. References to Adanson's scholastic progress, which appear in reports made by the prefect of the college, generally indicate that he was a well-liked student with personal attributes that would make for a good dragoman (*Père* Masson wrote of Adanson in 1746 that he: 'fera un très honnête homme... capable de rendre de très bons services').[9] However, in 1748 his academic talents were described as only 'médiocres', and in 1750 Adanson was on the verge of being expelled on account of his poor skills in Latin. Fortunately, the crisis was defused by the intervention of his instructors, who argued on Adanson's behalf that his grasp of Turkish was very good – it seems that he had even composed a Turkish word-list on his own initiative that was much admired. Further, it was also reported of Adanson that 'il écrit, il peint, il dessine à merveille',[10] and so he was saved from expulsion thanks to 'ces dispositions assez souvent indépendantes de la médiocrité du génie pour les lettres'.[11] Indeed, the records seem to indicate that Adanson had a good deal of talent in the areas that could be of practical use to a dragoman, so the fact that he came so close to expulsion because of his lesser achievements in Latin serves to underline the questionable priorities of the Jesuit attitude concerning the composition of a good education. In the end, however, the situation was resolved by simply sending Adanson to the Capuchin college at Pera, where he remained for two years, until July 1752, when he received his first assignment as a dragoman in Salonika, a port in north-eastern Greece.[12]

## Adanson Abroad: Official and Unofficial Sources

The word 'dragoman' is derived from the Turkish word meaning 'interpreter', which aptly sums up the essence (and limits) of the job. Despite the relatively low position of dragomen within the diplomatic hierarchy, as the only parties able to speak the languages of both the western and eastern powers, they found themselves privy to all the affairs of state. Dragomen were assigned to serve either the ambassador at Constantinople, or a consul or vice-consul at one of the French *échelles* such as those of Aleppo, Alexandria and Smyrna. Dragomen could be expected to act as interpreters, negotiators, unofficial 'observers' of the Ottoman government, spokespersons, law officials (for French subjects), official representatives and, perhaps as the end of a unusually successful career, as professors of Oriental languages in Paris.[13] The highest position that could be held by a dragoman was First Dragoman at Constantinople, serving the Ambassador of France in the Ottoman capital. What a dragoman could not do, however, was ever to become a vice-consul, consul, or ambassador himself; access to these prestigious positions was dictated by birth rather than ability.[14] The opening paragraph of a memo preserved in the Bibliothèque Nationale entitled *Nécessité d'encourager en France l'étude des Langues Orientales*[15] expresses the dimensions of the (pre-Revolutionary) glass ceiling very well:

> La littérature orientale est fort cultivée en Angleterre, en Allemagne, en Hollande, et dans d'autres pays européens, tandis qu'elle est fort négligée en France. Et la raison en est simple: c'est que dans ces États, les connoissances qu'on acquiert dans les langues orientales procurent des places honnorables, des emplois lucratifs, lorsqu'en France ces mêmes connoissances amènent tout au plus à obtenir une chaire au Collège Royal, une place d'interprète à la Bibliothèque Nationale, ou bien à vegeter toute sa vie, loin de sa Patrie, dans des places subalternes, qui procurent à peine la subsistance... Ce sont ces Agents les plus utiles et les moins payés.[16]

The attractions of the dragoman career (aside from the state-funded education) were mainly those benefits included in the *Capitulations* of 1740, the accords passed between the Ottoman Porte (the central Ottoman government in Constantinople) and France. According to this agreement, dragomen were exempt from Ottoman imposts and enjoyed reduced customs. Article 46 of this document also provided for the safety of dragomen, stating that they could not be imprisoned or reprimanded by Ottoman officials while carrying out their duties.[17] However, despite the potentially serious diplomatic consequences faced by the Ottomans should this accord be breached, the French dragomen could not afford to become complacent about their inviolability, as Adanson himself would experience. Furthermore, by the second half of the 18th century, the central authority within the Ottoman Empire had begun to weaken in favour of local Mameluke powers, and thus so too did the ability of France to enforce and rely upon the terms of the *Capitulations*.

The details of Adanson's career abroad are difficult to fix with complete certainty,[18] but it is clear that besides his initial stint in Salonika he also served in the *échelles* of Aleppo, Saida, Syrian Tripoli, Cairo, Alexandria, and Tunis. In Saida, Adanson became involved in a political gambit undertaken by its Pasha, which ended with tragic and lasting results for Adanson personally. In 1769 a Turkish ship was captured by a Maltese corsair, who also abducted the ship's crew to Malta to be sold as slaves. In retaliation, the Pasha of Saida seized two dragomen from the French consulate: First Dragoman Etienne Brüe, and Second Dragoman Jean-Baptiste Adanson.

In flagrant violation of the *Capitulations*, the two men were severely beaten by order of the Pasha, who used the threat of keeping them indefinitely as prisoners to force the French Consul of Saida to sign a document pledging to provide reimbursement for the lost ship and cargo, and to buy back the enslaved Turkish crew members from Malta. The consul immediately submitted to the Pasha's demands and effected the release of the unfortunate dragomen, but this event touched off a chain of political consequences that did not bode well for French prestige in the Levant. Although the demands to the Ottoman Porte that the Pasha of Saida be deposed were eventually met, there were so many delays (and other political machinations) in doing so that the incident became an indicator of the loss of French authority in its *échelles*, and particularly its ability to protect its citizens residing there. In August 1770 another Maltese raid resulted in the capture of several Turks from Beirut, and it was reported that 'on n'entendit qu'une voix dans les rues, qui disait que les coups de bâton ayant opéré l'année dernière le retour des esclaves ici, il fallait à la première occasion prendre quelques Français pour délivrer les esclaves qu'on pourrait venir faire sur cette côte'.[19]

However damaging the event may have been for the French nation, there can be no doubt that the experience was personally devastating for the dragomen who had been mistreated. There is little in the official record that describes Adanson's condition after his return. In a 1775 memo probably composed by M. de Saint-Priest, the Ambassador of France at the time, there is the following rather brief comment: 'Jean-Baptiste Adanson... Il a d'ailleurs le mérite d'avoir souffert pour le service à Seyde d'une bastonade dont il est demeuré un peu estropié. Il a été gratifié à cette occasion d'une pension de cinq cents livres.'[20] Fortunately, however, his story has also been conveyed to us via a different source, the travel account first published in 1799 by Sonnini de Manoncourt describing his visit to Egypt from 1777 to 1778.[21] Sonnini met Adanson in 1777 while the latter was serving as First Dragoman in Alexandria. Sonnini appears to have been impressed by Adanson's character and tragic history, both of which are commented upon in the following excerpt:

> Among the few Frenchmen who resided there [in the French factory[22] in Alexandria], and whose generous and obliging character can never be obliterated from my memory, I must distinguish a name dear to the Sciences, that of Adanson. The brother of the academician of Paris, devoted, from his youth up, to the study of the oriental languages, had for a long time fulfilled the delicate functions of an interpreter in the Levant. He had undergone, in Syria, one of those cruel inflictions, which are equally the reproach of the government which orders them, and of that which suffers them to pass unrevenged. The victim of his duty, he was likewise so of the detestable barbarity of a Pacha. Enjoined, in concert with his

colleague, and in the name of the French nation, to present well-founded remonstrances, they were both doomed, by order of the ferocious Mussulman, to the horrid punishment of the bastinado on the soles of the feet. The other interpreter expired under the hands of the executioner,[23] and Adanson, still more unfortunate perhaps, his feet mangled and exquisitely painful, almost entirely deprived of the power of walking, survived his execrable chastisement, and the affront offered to France, which her government left unpunished as well as that of the assassination of her consul at Alexandria.[24]

Apart from this act of reluctant heroism, however, Adanson's overall progress within the diplomatic ranks was not spectacular. For example, an assessment of Adanson's performance in 1766 states that he 'est peu versé dans les langues orientales, il a de l'honneur, mais peu de talens et de goût pour l'employ de drogman'.[25]

By contrast, the travellers who actually encountered Adanson and benefited from his aid seemed to form a much higher opinion of him. Sonnini's respect for Adanson is clear from the above-quoted passage and from his further remarks:

> Had I only to express my satisfaction in having received certain formal civilities, I should have dispensed with making a particular mention of these two interpreters [Adanson and his colleague, M. Augustus], without knowing whether they are still within reach of hearing this expression of my gratitude; but to them, and their illuminated complaisance, I stand indebted for the facility with which I was enabled to make my observations in countries of no easy access; and travellers will feelingly perceive the value of such encounters; for they know, as I do, how rare they are.[26]

Another French traveller, Claude Etienne Savary, had the opportunity to meet Adanson in Alexandria in 1777 at the beginning of his two-year journey around Egypt, the account of which was published in 1785 in Savary's *Lettres sur l'Égypte*. Although Savary's travel account does not emphasise his social contacts in general,[27] he does mention Adanson's name in the following episode which took place by the Sphinx on the Giza plateau:

> Tandis que nous admirions les merveilles de l'ancienne Egypte, & que M. Adanson, premier interprete du Roi à Alexandrie, étoit occupé à les dessiner, nous vîmes venir au galop dix Arabes la lance à la main. Ils s'approcherent à la portée du pistolet dans l'intention de nous attaquer, ou d'exiger un tribut. Nous étions armés de fusils & de pistolets & fort en état de les repousser; mais au premier feu, toute une tribu seroit venu fondre sur nous. Nous chargeâmes nos deux cheiks de leur parler. Ils leurs représenterent que nous étions leurs hôtes & qu'ils nous avoient pris sous leur sauve-garde. Ce seul mot les désarma, car ils respectent infiniment les droits de l'hospitalité. Ils descendirent de cheval & nous offrirent de nous accompagner par-tout où nous desirions aller.[28]

Even this brief encounter seems characteristic of the relentless anxiety that the local French merchants and consular personnel experienced during this time in Egypt. In the summer of 1777, the French consulate of Cairo, which had been the seat of the French consul in Egypt for more than a century, was abandoned due to the increasing security risks resulting from political instability in that city.[29] The consulate was transferred to Alexandria, despite the fact that its vice-consul, Boriès, had been murdered there the previous spring. Sonnini, who was present at the time of this transfer, wrote of the perpetual fear under which the French lived in Alexandria:

I was witness, one day, of the terror excited in the minds of the French belonging to the factory, from the idea only of a seditious tumult at Alexandria. A merchant arrived with intelligence that an European had killed a native of the country. The gate of the factory was instantly shut; the bales began to be put in motion to strengthen the bulwark; inquiries were made on board what vessel they could run for safety, by dropping from the windows, when, fortunately, information was received that one Mussulman had killed another.[30]

Despite these conditions of general unrest, on 2 January 1779 Adanson was entrusted with the care of a traveller of note, the Comte d'Entraigues, who was the nephew of M. de Saint-Priest.[31] Expressing a wish to see the lands of the Greek authors he professed to adore, d'Entraigues had accompanied his uncle as far as Constantinople as a starting place for an oriental sojourn. In Constantinople, however, he became involved with a Hungarian princess who happened also to be a favourite of the Sultan's. The two were seen together so often that a scandal began to brew. Saint-Priest quickely reminded his nephew that lingering in Constantinople was not part of his intended itinerary, and sent him on to Alexandria. As First Dragoman, and at the Comte's request, Adanson was placed at d'Entraigues' disposal to facilitate his excursions over the course of his two-month visit. The Comte soon perceived that Adanson was a skilled artist 'comme l'attestaient maintes planches où son art avait su fixer la forme et le teinte des plantes du désert et des poissons du Nil,'[32] and so charged Adanson with the job of drawing the monuments and antiquities that they would encounter during their travels.[33]

Despite the brevity of the Comte's stay, d'Entraigues' itinerary provides a list of at least some of the locations within Egypt at which Adanson was also certainly present. D'Entraigues' first excursion was to cross the desert from

Alexandria to the monastery of St. Macarius, where he claimed to have found precious Greek manuscripts in the monks' library. Heading back into the Delta, they stopped at Rosetta to embark on the Nile. They then proceeded to Cairo by boat, where they stayed in the *oquelle* of the French merchants who had chosen to remain, despite the recent removal of the consulate and its protection. D'Entraigues' next excursion took them to Suez, from where they climbed Mt. Horeb and Mt. Sinai. Although d'Entraigues had intended to journey as far south as the cataracts, he was prevented from doing so by Hassan bey Djeddauoui, a local power who was encamped near Edfu. The Comte claimed to make it as far south as Thebes, but Auriant suggests it is more likely that he could not proceed further than Antinoe.[34] Disappointed with his trip, d'Entraigues returned to Constantinople at the end of February, carrying with him a portfolio of Adanson's drawings to supplement his own notebooks of observations.[35]

Beyond the bare facts of Adanson's travel within Egypt, d'Entraigues' account also reveals a certain amount of insight into Adanson's personal interests and skills. Despite the mediocre assessments that seemed to hound him from his school days and on into his career, Adanson was nevertheless designated as the guide for the ambassador's own nephew who, in his turn, was favorably impressed with Adanson's abilities, not only as an able official, but as an informed artist. Thus, one begins to see the manner in which Adanson employed those artistic skills that were always acknowledged as remarkable: d'Entraigues' account mentions Adanson's depictions of flora and fauna (quoted above), as well as the Egyptian antiquities he was urged to record for the Comte.

Although unlike his brother, Michel, Adanson was no expert in natural history, there is evidence from another contemporary traveller, C.-F. Volney, which demonstrates his scholarly interest in this topic. In his *Voyage en Syrie et en Égypte pendant les années 1783, 1784 & 1785*, a sober account of the physical and political states of those two nations, Volney mentions Adanson within the context of an exposition on the natural history of Syria:

> Enfin, il ne faut pas oublier d'observer que l'espece de colibri [hummingbird] existe dans le territoire de Saide. M. J.B. Adanson, ci devant interprete en cette ville, qui cultive l'Histoire Naturelle avec autant de goût que de succès, en a trouvé un dont il a fait présent à son frere l'Academicien.[36]

Indeed, there is evidence from Adanson's own writings that he collected specimens of local flora and fauna, which he not only drew but also carefully preserved, in the fashion of the times, in a curio cabinet. For example, in Adanson's treatise on natural history,[37] a margin note corresponding to his entry on the 'remora' contains this comment: 'J'en ai disséqué deux qui se trouvent dans mon Cabinet d'histoire naturelle. On en verra la description dans mes ouvrages. On nomme ce poisson à Alexandrie *Canule*.'[38] It is interesting to learn that Adanson, who never returned to Europe, was nonetheless fully participating in the intellectual trend for classification that was rampant there at that time. Thus, even a slight comment from a traveller who benefited from Adanson's friendship discloses still more about his talents and interests, and helps to create a fuller context in which to evaluate Adanson's approach to Egyptian antiquities, which is the focus of the Adanson manuscript to be discussed below.

## Adanson's Scholarship regarding Egyptian Antiquities

According to Hamy, in 1795 while serving as First Dragoman-Chancellor in Tunis, Adanson received a letter from the Comité d'instruction de la Convention nationale inviting him

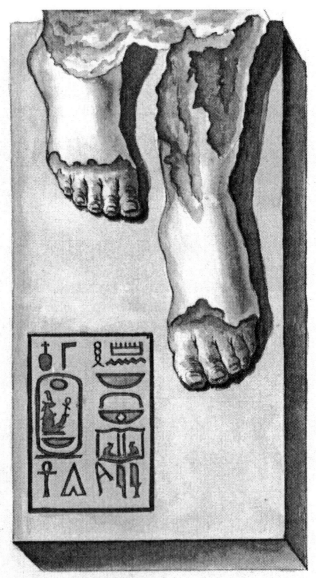

**Figure 5:** Fragment of an Egyptian statue from *Origine des Egyptiens et de leur écriture symbolique*, Jean-Baptiste Adanson (Plate 17).

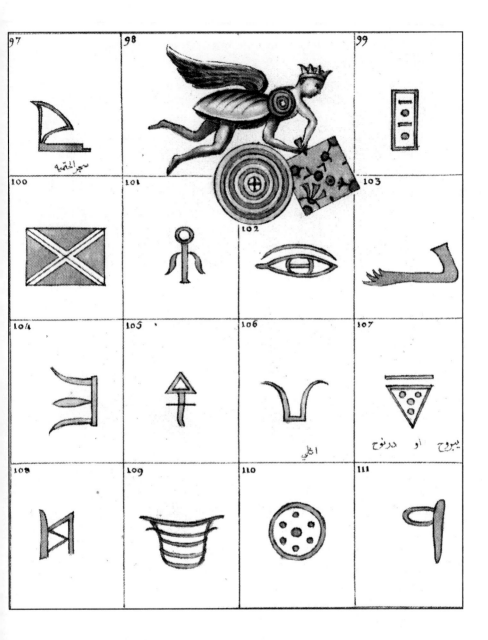

**Figure 6:** Hieroglyphic inscriptions from *Origine des Egyptiens et de leur écriture symbolique,* Jean-Baptiste Adanson (Plate 24).

to submit 'les divers ouvrages qu'il a composés pendant son séjour en Égypte et en Barbarie, afin qu'on puisse les faire imprimer et en faire jouir le public.'[39] Unfortunately, this work was not completed before Adanson's death in 1804, after which his manuscript disappeared and was presumed lost. However, a collection turned up in 1999. It is now in the Milton S. Eisenhower Library at the Johns Hopkins University, and appears to be the missing work. The subject matter of this manuscript is ancient Egypt, as represented by Adanson's illustrations and accompanying textual commentary on Egyptian antiquities, monuments, and hieroglyphic inscriptions, to which he had direct access during the ten years he spent in Egypt (c. 1775–1785). The collection comprises four main notebooks,[40] containing more than 50 pages of ink-washed drawings as well as approximately 50 pages of text executed in careful handwriting. In addition, there is a fifth notebook which seems to contain some of Adanson's notes, together with rough sketches of hieroglyphs. The collection is completed by several loose sheets of paper featuring more sketches of hieroglyphs along with their purported translations.

From his introductory remarks it is clear that Adanson believed that he could offer two categories of useful and unique information to the European public. Firstly, there was the unquestionable value of his illustrations of genuine Egyptian objects since, in Adanson's day, reliable copies of Egyptian artifacts were still difficult to obtain. Secondly, he offered a translation of an Arabic manuscript, reportedly acquired by Adanson in Egypt, which claimed to furnish the translations of 331 hieroglyphs. When one considers the limited access that Europe had to Egypt and its antiquities at that time (a good few years before Napoleon Bonaparte invaded Egypt and opened its borders to Western contemplation), these offerings can be understood to have been quite valuable to European scholarship and imagination. In order to evaluate Adanson's treatise, with particular reference to its place within the development of Egyptology as an academic field of study, the content of his manuscript will here be divided into three components for consideration: his drawings of antiquities bearing hieroglyphic inscriptions, his textual commentary, and his presentation of the Arabic manuscript purporting to give the translations of Egyptian hieroglyphs.

As mentioned above, Adanson's treatise included more than 50 pages of depictions of antiquities; these feature objects ranging from a standing obelisk, to inscribed statue fragments, to magical amulets. The common element among the illustrations Adanson chose to include seems to be the presence of hieroglyphic inscriptions which, as will be seen below, held a particular interest for him. The majority of the antiquities drawn by Adanson still exist and have been identified in various museum collections, which allows the accuracy of his depictions to be tested against the objects as they appear today. Through such comparisons, one can evaluate Adanson's attitude towards his subjects.

For example, Adanson's rendering of a statue base from the reign of Amenhotep III is shown in Figure 5; he reports that it was 'Found near Alexandria; Drawn on the spot by J.B.$^{te}$ Adanson'.[41] A comparison of Adanson's version of the inscription with a photograph of that object as it currently appears in the Musée Calvet in Avignon quickly demonstrates that Adanson was making an earnest attempt to convey the hieroglyphs accurately. Although he did miss the *pt*-sign that forms the top of the text box, and misunderstood some of the details of the hieroglyphs, these can be interpreted as the mistakes made by an artist who was unfamiliar with the nuances of the Egyptian script; they certainly are not so far from reality as to suggest an intentional deception designed to produce hieroglyphs in accordance with a preconceived notion of how they ought to appear. In fact, Adanson's version of the statue base inscription is quite readable, and can be translated as

'Nebmaatre, the good god, given life, beloved of Hemen, lord of the Sed-festival'.[42] Despite the quality of his copy of the inscription however, it must be conceded that in this example, as in others, Adanson did not succeed in capturing the general style of Egyptian art, a problem that also bedevilled his predecessors and contemporaries.

It is important to examine the extent (if any) to which Adanson projected his own preconceived expectations onto his renderings of Egyptian hieroglyphs and antiquities, not only to assess how confidently his illustrations can be used for purely Egyptological information, but also as an indicator of the transitional state of Egyptology at the time Adanson was working. European conceptions of (and fascination with) ancient Egypt were constructed long after the civilisation itself had become extinct. During the Renaissance, for example, Egypt was reborn in the minds of Europeans as a fount of precious ancient wisdom, a perception supported by Neoplatonist writings. Indeed, this view that Egyptian hieroglyphs contained esoteric knowledge would persist in varying degrees until the decipherment of the Rosetta Stone and the subsequent translation of Egyptian texts (which were often revealed to be concerned with more mundane matters).

Since the decipherment would not take place until 1822, in Adanson's time ancient Egypt could still be made to conform to whatever modern European society wished it to be. Even during the more staid Enlightenment, fanciful ideas of ancient Egypt still featured in some of the intellectual debates of the day, and Adanson was certainly not immune to this influence, as his textual commentary demonstrates (below). Nevertheless, Adanson's frequent assertion, that various items were 'copied exactly' or drawn 'on site', combined with the generally realistic effect of his finished illustrations, do strongly suggest that he strove to achieve objective accuracy.

But if Adanson's illustrations of Egyptian antiquities

seem to be a straightforward collection of useful data, an examination of Adanson's textual commentary exhibits the conflicting influences that characterised pre-decipherment studies of Egyptological subjects. Adanson covers a bewildering range of topics in his text: the origins of the Egyptians and their writing system, the symbolic significance of Horus and Harpocrates, the Egyptians' symbolic use of live animals, and so on. Furthermore, there is no obvious relationship between these topics and the illustrations which they supposedly explicate, nor does Adanson make any such connection explicit by referring the reader to the images in his plates.

This nearly total conceptual rift between Adanson's text and his images is partially explained by the fact that, in most cases, the ideas that Adanson was expressing were not the results of his own observations, but the ideas of an 18th century scholar, Noël Antoine Pluche. In fact, the majority of Adanson's text is taken word for word from Pluche's book, *Histoire du Ciel*,[43] which employed an inaccurate vision of ancient Egypt, moulded to support Pluche's larger goal of using 'science' to prove the legitimacy of the Old Testament account of the creation of the universe.

Although it is now obvious that Pluche's conclusions about ancient Egypt were completely unfounded, it must be remembered that there was no evidence available at that time to disprove them; indeed, Pluche's theories were essentially derived from Neoplatonic scholarship on the subject, which would continue to be the ultimate authority on ancient Egypt until the decipherment of hieroglyphs. For example, Pluche (and through him, Adanson) characterised the origin of the Egyptian writing system as a collection of public notices in an allegorical language established by astronomer-priests for the purpose of warning the populace of the rise and fall of the Nile's inundation, and informing them of which agricultural activities should accompany each season.[44] As Adanson himself noted,[45] Pluche never had occasion to visit Egypt, and instead had relied on the same limited (and faulty) sources that were available to Renaissance scholars, such as the *Mensa Isaica*, and the works of Diodorus of Sicily and Herodotus.

While it is not surprising that Pluche's interpretations are so decidedly erroneous, Adanson's reliance on Pluche's work is at first somewhat puzzling, particularly in light of his obvious advantage in having access to the ancient Egyptian monuments themselves. However, one must appreciate that Adanson's situation was actually quite similar to that of Pluche and other scholars who speculated at will on the nature of ancient Egypt. Despite Adanson's access to the primary sources, he still lacked the interpretive tools with which to understand them. Given this state of knowledge, the most that Adanson could reasonably achieve was to use his (relatively) unbiased observations of the evidence to express some doubts about Pluche's interpretations – which he did, to some extent. For instance, Adanson noted that the hieroglyphs described by Pluche 'are hardly in accordance with those which are seen on the obelisks which still exist in Egypt, and in so many other monuments which I have seen during a residence of ten years in this region...'[46]

However, he hastened to account for this discrepancy by suggesting that the obelisks he could observe in Egypt were of greater antiquity than were the sources available to Pluche, thereby avoiding a direct refutation of Pluche's account.[47] Whereas during the 18th century other fields of study (such as geology and natural history) could flourish almost instantly under the scientific processes of observation, Egyptology required the additional spur of comprehension of the ancient texts before it could emerge as an objective field. This was Adanson's situation, and his response to it provides a study in microcosm of the larger intellectual context into which Egyptology was eventually born; while Adanson's illustrations display an Enlightenment-era desire for objectivity, his interpretive texts still reflect the centuries-old limits that could not be overcome until the decipherment.

Finally, there is the third aspect of Adanson's work to consider: his presentation of a manuscript which claimed to offer the key for the correct interpretation of 331 Egyptian hieroglyphs. This subject takes up nearly all of his *Second cahier*. He begins with a short explanatory text and follows with plates showing the so-called hieroglyphs organised into numbered grids (see Fig. 6). Adanson entitled this section 'Index or Table of the Symbolical Figures',[48] and made the following remarks:

> Copied exactly from an Ancient Manuscript in the Greek Idiom Found in the Ruins of Thebes in upper Egypt and Translated into Arabic by a Copt. This is an accurate Extract which I made in Egypt, from the Original Manuscript which a Friend Obtained for me... I do not at all claim to give the slightest certainty on the explanation of all the Symbolical figures and Mystical characters contained in this Manuscript. However, a large part seems sufficiently analogous to the Hieroglyphs of various obelisks and other Egyptian monuments of which I give here the most precise plans, which I Copied from the actual objects which still exist in Egypt. The Public will not be ungrateful for my research of this kind nor for the pleasure which I take at the same time in communicating to it this little Manuscript, which will be able perhaps to give some insights into the Mystical Science of the Egyptians which has been explained in various ways, by ancient authors as well as Modern ones. The historian Pluche seemed to me to have reasoned out this still obscure Science with a good deal of Judgment. I even presume, that he could have pushed his knowledge further, if he had travelled to Egypt, where he could have observed all the ancient monuments which still exist there, and particularly those of the ancient City of Thebes in the Saidi or upper Egypt.[49]

Adanson did not seem to have finished his final translation of this Arabic document. Although he provided the images of the hieroglyphs in a finished form in the *Second cahier*, the corresponding translations that one would expect are absent. Fortunately, however, his initial efforts are included among the rough drafts that are part of the larger collection, so it is possible to reconstruct the majority of the Arabic manuscript's hieroglyphic decipherments. For the most part, these decipherments consist of nouns or adjectives which are grouped, as Adanson noted, into the following categories: religious dogmas of the ancient Egyptians, astronomy, medicine, plants and minerals.[50]

Adanson's occasional retention of the original Arabic in his notes and in the grid suggests that he truly was in possession of an actual document, and that this portion of his treatise is not simply fantasy. However, since the source of this document is still unknown to me, the accuracy of Adanson's copies of the manuscript's images cannot be evaluated. Assuming that Adanson continued to employ the same care with which he made his his illustrations of inscriptions, one can say that among the 331 images that the Arabic manuscript apparently provided, there are some that do bear a resemblance to those in the ancient Egyptian script (*see* Figure 6, grid nos. 102 and 103, for example). However, there are many more that can only be described as fantastic (*see* Figure 6, grid. no. 98 for the most outlandish of the collection).

Despite Adanson's comment (quoted above) that he was comfortable with the legitimacy of these hieroglyphs, finding them 'sufficiently analogous' to the hieroglyphs he observed on Egyptian monuments, it is telling that Adanson never made any attempt to utilise this information to provide translations for the ancient Egyptian inscriptions which he had personally copied. One might argue that Adanson had fully intended to undertake this task, and would have done so had he not passed away prematurely;

however, I suggest that it is more likely that the inevitable difficulties Adanson would have encountered in applying the false hieroglyphs from the Arabic manuscript to the more realistic signs in his own copies was the real impediment to the completion of his publication.

It is clear, then, that Champollion's place in history as the first to decipher hieroglyphs faces no threat from Adanson. The true value of Adanson's work lies in the unique perspective it provides by illustrating a remarkable transitional moment in the history of Egyptology: the moment when it teetered on the line between scientific endeavor and nearly limitless speculation. The extended education that Adanson received in his youth, his friendship with erudite scholars visiting Egypt, his interest in natural history, and his direct access to the monuments of Egyptian antiquity are all factors which seem to have inclined Adanson to a typically 18th-century belief in the value of objectivity. Nevertheless, when confronted with the task of interpreting his scientific data, Adanson could not completely resist the allure of 'Mystical characters' which could be revealed by a single fortuitously discovered manuscript. Thus, we can perceive Adanson also hesitating on an important threshold, unable in good conscience to fully embrace the spurious manuscript, but nonetheless unwilling to dismiss it entirely as a fraud.

To conclude: a truly scholarly approach to the study of ancient Egypt could not come into existence until the decipherment of hieroglyphs had taken place: the effect of the ability to read hieroglyphic texts was to provide irrefutable (and immutable) evidence of the actual life-ways of the ancient civilisation, thereby creating a divide between the Egypt of the scholars and the Egypt of the dreamers. During Adanson's lifetime the processes were begun – but not yet completed – that would divide interest in Egypt into the separate spheres of Egyptology (scientific method) and Egyptomania[51] (without scientific method). Thanks to our knowledge of some of the influential events in Adanson's life abroad and the works that resulted from his tenure in Egypt, we have an unusual record of how those processes were experienced by a man on the cusp of a new era of scholarship.

REFERENCES

1. The biographical information in this paragraph comes from the followingfour sources: Anne Mézin, *Les Consuls de France au Siècle des Lumières (1715-1792)*, (Paris: Ministère des affaires étrangères. Direction des archives et de la documentation, 1998), pp. 85-87; Antoine Gautier and Marie de Testa, 'Quelques Dynasties de Drogmans,' *Revue d'histoire diplomatique* 105 (1991): pp. 41-44; Henri Guérin, 'Adanson (Jean-Baptiste),' in *Dictionnaire de biographie française*, ed. J. Balteau, et al., (Paris: Letouzey et Ané, 1933-), p. 506; and Ernest-Théodore Hamy, 'Un égyptologue oublié: Jean-Baptiste Adanson (1732-1804),' *Comptes rendus des séances de l'année 1899 (Académie des Inscriptions et Belles-Lettres)*, 27, 4th Series (1899): 738-746. This last source was also published in: idem, *Les débuts de Lamarck*, Bibliothèque d'histoire scientifique, vol. 2, (Paris: E. Guilmoto, 1908): pp. 37-46.
2. Michel Adanson went on to become a famous naturalist, and author of the influential works *L'Histoire naturelle du Sénégal* (1757) and *Les familles des plantes* (1763).
3. An *échelle* was a port city frequented by European ships and, by extension, the ports or other politically or economically important cities where there existed small colonies containing a permanent populace of European traders, as well as artisans, merchants, civil servants, etc. (Mèzin, *Les Consuls de France*, p. 29, n. 2).
4. Gustave Dupont-Ferrier gave a comprehensive history of the school in the following two articles: 'Les Jeunes de Langues ou 'arméniens' à

Louis-le-Grand,' *Revue des études arméniennes* 2 (1922): 189-232; and 'Les Jeunes de Langues ou 'arméniens' à Louis-le-Grand', *Revue des études arméniennes* 3 (1923): 9-46; the information in these articles is repeated in the same author's 'Appendice M' in *Collège de Clermont au Lycée Louis-le-Grand* (1563-1920), Vol. III, (Paris: Boccard, 1925): 349-448.

5. See Marie de Testa and Antoine Gautier, 'Les drogmans au service de la France au Levant', *Revue d'histoire diplomatique* 105 (1991), pp. 7-38.
6. Marseille had an interest in increasing French influence in the Levant in the hope that a commercial trade monopoly might be created for its own port.
7. 'I am obliged to say... that no nation is so badly served in dragomen than ours' [translation Kimpton]. Dupont-Ferrier, 'Jeunes de langues,' 1922, p. 196.
8. Levantine French students were usually the sons of dragomen in the field who were themselves former *jeunes de langues*. The preference for admitting the children of current dragomen working abroad was a benefit designed to instill further loyalty to the French state, and sometimes created rather influential dragoman 'dynasties'. Adanson certainly took advantage of the system: all three of his sons attended Louis-le-Grand (Charles-Louis, 1775-1784; Pierre-Flavien, 1783-1785; Alexandre-Victor, 1789-1797).
9. '[He] will make a very honest man... capable of rendering very good services' [translation Kimpton]. From the *Archives du Ministère des affaires étrangères* as cited by Hamy, 'Un égyptologue oublié,' pp. 740-741.
10. 'he writes, paints and draws excellently' [translation Kimpton]. Hamy, 'Un égyptologue oublié', p. 741.
11. 'these dispositions quite often independent of the mediocrity of genius for letters' [translation Kimpton]. ibid.
12. Gautier and de Testa, 'Dynasties de drogmans', p. 41.
13. de Testa and Gautier, 'Drogmans au service de la France,' pp. 7-13.
14. This situation would change after the French Revolution, and already

in 1776 the post of 'chancellor' was eliminated from some *échelles* and the duties devolved upon dragomen nominated by their consuls.

15. 'The Necessity of encouraging the study of Oriental Languages in France' [translation mine]. This document and others pertaining to the lives of dragomen during the 18th century can be found in Henri Cordier, 'Un interprète du Général Brune et la fin de l'École des jeunes de langues', *Mémoires de l'Institut national de France, Académie des Inscriptions et Belles-Lettres* 38, 2nd Part (1911): 267-350.
16. 'Oriental literature is strongly cultivated in England, Germany, Holland, and in other European countries, whereas it is very much neglected in France. And the reason for it is simple: it is because in these Nations, the knowledge that one acquires in the oriental languages obtains honorable positions, lucrative employment, while in France, this same knowledge leads at best to obtaining a chair at the College Royal, a position as interpreter at the Bibliothèque Nationale, or else to vegetate all one's life, far from his Homeland, in subordinate positions, which scarcely procure a livelihood ... These Agents are the most useful and the least paid' [translation Kimpton].
17. de Testa and Gautier, 'Drogmans au service de la France', p. 19.
18. Differing versions of Adanson's career may be found in Hamy, 'Un égyptologue oublié', pp. 738-746; Gautier and de Testa, 'Dynasties de drogmans', pp. 41-44; and Mézin, *Les consuls de France*, pp. 85-87; all of which, nevertheless, rely on the same primary sources, mainly the diplomatic correspondence and other documents preserved in the *Archives nationales* and the archives of the *Chambre de commerce de Marseille*. At the time of publication, the author has not yet had the opportunity to examine these documents personally.
19. 'Only one voice was heard in the streets, which was saying that the blows of the rod, having brought about the return of the slaves here last year, it was necessary to capture a few Frenchmen at the first opportunity in order to release the slaves...' [translation Kimpton]. From a letter written by M. de Saint-Priest, Royal Ambassador at Constantinople, cited in François Charles-Roux, *Les échelles de Syrie et de Palestine au XVIIIe Siècle* (Paris: Paul Geuthner 1928), p. 90. See ibid., pp. 88-90 for more about the political ramifications of the incident.
20. 'Jean-Baptiste Adanson... Moreover he has the merit of having suffered for the department at Saida from a bastinadoing from which he remained slightly crippled. He was presented at that time with a pension of five

hundred livres' [translation Kimpton]: de Testa and Gautier, 'Drogmans au service de la France', p.19, citing document AEB III 244, fo 133 and following from the Archives nationales.

21. See Sonnini de Manoncourt, *Travels in Upper and Lower Egypt*, trans. H. Hunt, (London: J. Stockdale, 1779), especially vol. I, chap 12 (p. 182 ff.).

22. The term 'factory' (also *oquelle, auquelle* or *okel* in French documents) refers to a group of buildings arranged in a rectangle facing a central courtyard with the only access point being a central gate which can be secured; the bottom floor contains warehouses while the upper floor has residence areas (after Mézin, *Les consuls de France*, p. 30). For a labelled plan of the factory at Alexandria (c. 1730) see Raoul Clément, *Les Français d'Égypte aux XVIIe et XVIIIe siècles*, (Cairo: Institut français d'archéologie orientale, 1960), p. 154.

23. Etienne Brüe did not die during this ordeal, but survived and was compensated with 800 livres. He died in 1775 in Saida following a long illness. See Gautier and de Testa, 'Dynasties de drogmans', pp. 47-48.

24. Sonnini, *Travels*, pp. 184-185.

25. 'little versed in the oriental languages, he has the honour, but little talent or liking for the employment of dragoman' [translation Kimpton]. Gautier and de Testa, 'Dynasties de drogmans', p. 42.

26. Sonnini, *Travels*, pp. 185-186.

27. Savary's purpose in visiting Egypt was to perfect his Arabic language skills, so it may be that he did not spend a great deal of time with his French compatriots (see Clément, *Français d'Égypte*, pp. 240-244).

28. 'While we were admiring the marvels of ancient Egypt, and M. Adanson, first interpreter of the King at Alexandria, was busy drawing them, we saw ten Arabs coming at a gallop, spears in hand. They drew up to pistol range with the intention of attacking us, or exacting a tribute. We were armed with guns and pistols and very much in a condition to push them back, but at the first fire, an entire tribe came to swoop down on us. We instructed our two sheiks to speak to them. They represented to them that we were their guests and that they had taken us under their protection. This single word disarmed them, for they infinitely respect the laws of hospitality. They dismounted and offered to accompany us wherever we wished to go' [translation

Kimpton]. C. E. Savary, *Lettres sur l'Égypte* (Paris: Onfroi, 1785), pp. 220-221.

29. This action was taken in accordance with the recommendation of the royalty-appointed Inspecteur général des Échelles, Baron de Tott. For an overview of the problems in Cairo (from a French perspective) between 1772 and 1792 see Clément, *Français d'Égypte*, pp. 217–238.
30. Sonnini, *Travels*, p. 190.
31. For the journey of d'Entraigues see Auriant, 'Le voyage orientale de M. d'Entraigues', *Revue bleue* 68 (1930), pp. 293-296; Clément, *Français d'Égypte,* pp. 247-251; and Jean-Marie Carré, *Voyageurs et écrivains français en Égypte I: de début à la fin de la domination turque (1517-1840)*, (Cairo: Institut français d'archéologie orientale, 1932), pp. 105-107. Carré reported that d'Entraigues' own account of his journey was then in the holdings of the Bibliothèque de Dijon.
32. Auriant, 'Voyage orientale', p. 294.
33. This circumstance probably explains Adanson's contact with Savary at Giza (mentioned above). Savary had noted on an earlier visit to the pyramids: 'J'arrive d'un second où s'est trouvé le Comte d'Antragues, que le desir de s'instruire a conduit en Egypte. Ce Seigneur François joint aux qualités aimables, beaucoup d'esprit & de connoissances' (Savary, *Lettres*, p. 160).
34. Auriant, 'Voyage orientale', p. 295.
35. ibid.
36. 'Finally, we must not forget to observe that the species of hummingbird exists in the territory of Saida. M. J.B. Adanson, previously an interpreter in this town, who cultivates Natural History with equal taste and success, found one there of which he made a present to his brother the Academician' [translation Kimpton]. Constantin-François Volney, *Voyage en Syrie et en Égypte pendant les années 1783, 1784 & 1785*, (Paris: Volland and Desenne,1789), p. 193, footnote (b). Although Volney mentions Adanson's name in reference to Saida, it is likely that the two actually met when Volney arrived in Alexandria in 1783.
37. Jean-Baptiste Adanson, *Histoire naturelle*, in *Adanson Collection*, Ms. 396, Special Collections, Milton S. Eisenhower Library, The Johns Hopkins University. This large, handwritten treatise appears to be the

result of many years of work, judging by the frequent additions, corrections, and margin notes.
38. 'I have dissected two of them which are in my Cabinet of natural history. The description of it shall be seen in my works. In Alexandria this fish is called *Canule'* [translation Kimpton].
39. 'the various works that he composed during his stay in Egypt and Barbary, so as to be able to have them published and let the public enjoy them' [translation Kimpton]. Hamy, 'Un égyptologue oublié', p. 745.
40. The notebooks are entitled: *Premier cahier concernant les hiéroglyphes; Second cahier concernant les hiéroglyphes; Troisieme cahier concernant les amulettes et autres* (which contains only illustrations); and finally, *Cahier concernant les figures symboliques des anciens Egyptiens.*
41. Adanson Collection, *Premier cahier*, plate 17, translation mine.
42. See Marie-Pierre Foissy-Aufrère, *Egypte & Provence*, (Avignon: Fondation du Muséum Calvet, 1985), inv. no. A38.
43. Noël Antoine Pluche (trans. J.B. DeFreval), *The History of the heavens, considered according to the notions of the poets and philosophers, compared with the doctrines of Moses* (London: J. Osborn, 1740), especially pp. 31-92.
44. Adanson Collection, *Cahier concernant les figures symboliques*; Pluche, *History of the Heavens*, pp. 61-65.
45. Adanson Collection, *Second cahier*.
46. Adanson Collection, *Premier cahier*, translation Kimpton.
47. ibid.
48. Adanson Collection, *Second cahier*, translation Kimpton.
49. ibid.
50. ibid.
51. 'Egyptomania' is defined here not only as a trend in decorativearts and architecture, but as the wider phenomenon of the European fascination with Egypt that tended to be fuelled more by imagination than by empirical observation.

# BIBLIOGRAPHY

*Adanson Collection* Ms. 396. Special Collections, Milton S. Eisenhower Library, The Johns Hopkins University.

Auriant, 'Le voyage orientale de M. d'Entraigues'. *Revue bleue*, 68, no. 10 (17 May 1930), pp. 293-296.

Carré, Jean-Marie, *Voyageurs et écrivains français en Égypte I: de début à la fin de la domination turque (1517-1840)*, Recherches d'archéologie, de philologie et d'histoire, vol. IV. Cairo: Institut français d'archéologie orientale, 1932.

Charles-Roux, François, *Les échelles de Syrie et de Palestine au XVIIIe Siècle*. Bibliothèque archéologique et historique, vol. X. Paris: Paul Geuthner, 1928.

Clément, Raoul, *Les Français d'Égypte aux XVIIe et XVIIIe siècles*. Recherches d'archéologie, de philologie et d'histoire, vol. 10. Cairo: Institut français d'archéologie orientale, 1960.

Cordier, Henri, 'Un interprète du Général Brune et la fin de l'école des jeunes de langues.' *Mémoires de l'Institut national de France, Académie des Inscriptions et Belles-Lettres*, 38, 2nd Part (1911), pp. 267-350.

Dupont-Ferrier, Gustave, 'Les jeunes de langues ou "arméniens' à Louis-le-Grand.' *Revue des études arméniennes*, 2 (1922), pp. 189-232.

—, 'Les jeunes de langues ou "arméniens" à Louis-le-Grand.' *Revue des études arméniennes*, 3 (1923), pp. 9-46.

—, *Du collège de Clemont au Lycée Louis-le-Grand (1563-1920), Vol. 3*. Paris: Boccard, 1925.

De Testa, Marie and Antoine Gautier. 'Les drogmans au service de la France au Levant.' *Revue d'histoire diplomatique*, 105 (1991), pp. 7-38.

Foissy-Aufrère, Marie-Pierre (publication director), *Egypte & Provence*. Avignon: Fondation du Muséum Calvet, 1985.

Gautier, Antoine and Marie de Testa, 'Quelques dynasties de drogmans.' *Revue d'histoire diplomatique*, 105 (1991), pp. 39-102.

Guérin, Henri, 'Adanson (Jean-Baptiste).' In *Dictionnaire de biographie française*, edited by J. Balteau, et al., p. 506. Paris: Letouzey et Ané, 1933-.

Hamy, Ernest-Théodore, 'Un égyptologue oublié: Jean-Baptiste Adanson (1732-1804).' *Comptes rendus des séances de l'année 1899 (Académie des Inscriptions et Belles-Lettres)*, 27, 4th Series (1899): 738-746.

—, *Les débuts de Lamarck*, Bibliothèque d'histoire scientifique, vol. II. Paris: E. Guilmoto, 1908.

Mézin, Anne, *Les consuls de France au siècle des lumières (1715-1792)*. Paris: Ministère des affaires étrangères, Direction des archives et de la documentation, 1998.

Pluche, Noël Antoine, *The history of the heavens, considered according to the notions of the poets and philosophers, compared with the doctrines of Moses*. Trans. J.B. DeFreval. London: J. Osborn, 1740.

Savary, Claude Etienne, *Lettres sur l'Égypte*. Paris: Onfroi, 1785.

Sonnini de Manoncourt, *Travels in Upper and Lower Egypt*. 3 Vols. Translated by Henry Hunt. London: J. Stockdale, 1799.

Volney, Constantin-François. *Voyage en Syrie et en Égypte pendant les années 1783, 1784 & 1785*. 2 vols, Paris: Volland and Desenne, 1789.

## Chapter Four
# The journey of the comte de Forbin in the Near East and Egypt, 1817-1818

## Pascale Linant de Bellefonds

On 3 October 1815, the Baron Dominique Vivant Denon, Director-General of Museums, handed in his resignation to the king of France, Louis XVIII, on account of his 'advancing age' (he was 68) and 'ill health'. A few months later, on 16 June 1816, he was succeeded by the Comte Auguste de Forbin, who was to retain this prestigious post until his death in 1841.

Strangely enough, the two men had several things in common: they were both trained artists and each, in his own way, had followed the Napoleonic campaigns, stayed for a while in Italy, travelled to Egypt and been successful in every field, particularly with the ladies. However, we now realise the degree to which Vivant Denon's personality overshadowed that of his successor, who nevertheless played a central part in the artistic life of the first half of the 19th century.

Louis Nicolas Philippe Auguste, Comte de Forbin, was almost 40 when he was appointed Director of the royal

Museums. He was a handsome man (Fig. 7), described by the *Quarterly Review* as 'the most dapper, best dressed gentleman in Paris',[1] a man about whom Chateaubriand wrote in his *Mémoires d'Outre-Tombe* that he 'held the hearts of princesses in his powerful hands'. This may well reveal a hint of jealousy, since Forbin had been passionately in love with Juliette Récamier and the rival of Chateaubriand for her heart. It was for the same Madame Récamier that, in 1814, Forbin nearly fought a duel with Benjamin Constant. This was how one of his contemporaries, the Vicomte Siméon, described him: 'Une taille élevée, une tournure élégante et noble, des traits réguliers et qui rappelaient les belles têtes du siècle de Louis XV, en faisaient ce qu'on eût appelé dans l'ancienne cour un gentilhomme accompli.'[2]

Auguste de Forbin belonged to the oldest nobility in Provence. Born on 19 August 1777 in the château of La Roque-d'Anthéron (Bouches-du-Rhône), he was soon affected by the turmoil of the Revolution: his family's fortune was confiscated and, in 1793, his father and uncle died on the scaffold in Lyon. Now orphaned, he was, as he said himself, 'saved from destitution' by the painter and engraver Jean-Jacques de Boissieu, who became his teacher in Lyon. Then in 1795, together with his friend Marius Granet, he joined David's studio in Paris. He devoted himself to chivalresque scenes and interiors, showing a taste for the dramatic effects and contrasts of light that feature in all his later works.

In 1799, during the Directoire period, he was enlisted in the army for two years, and then went to Italy to study painting. In 1805 he was appointed chamberlain to Princess Pauline Borghèse, Bonaparte's sister. Malicious gossips said that he was also her lover. Later he took part, as a field officer, in the Napoleonic campaigns in Portugal, Spain and Austria and was awarded the Légion d'Honneur. After Wagram, in 1809, he went once again to Italy to indulge his passion for the arts. It was in Rome, in 1813, that he met Madame Récamier and became a regular of her salon. He returned to Paris in 1814, two years before being appointed Director-General of Museums and three years before setting out for the Near East and Egypt.

To understand better the context and the state of mind in which Forbin accomplished his journey to the Levant in 1817, a few words are needed about the particular circumstances in which he took up his post of Director of Museums. This was the time of the Restoration, and Vivant Denon had just solved the thorny problem of returning the works of art accumulated since 1797 as spoils of war. Denon fought inch by inch with the foreign powers' envoys but, despite his courage and determination, over 5000 works were forcibly taken from the Louvre, half of which were antiquities. The largest museum in Europe was bled dry. Denon wrote: 'Des circonstances inouïes avaient élevé un monument immense; des circonstances non moins extraordinaires viennent de le renverser. Il avait fallu vaincre l'Europe pour former ce trophée; il a fallu que l'Europe se rassemble pour le détruire.'[3]

Forbin's first task was to reorganise and redesign the museum. It must be said that the feeling of despoliation provoked by the return of the works of arts was aggravated by the fact that, for the past few years, the British had been engaged in massive archaeological trawls in Greece, and in August 1816 the Elgin Marbles were transported to the British Museum. So France had to make up for lost time and it was with this objective that, in 1817, Forbin led his expedition to the Near East. Two people seem to have played a decisive role in his decision to undertake this trip: Vivant Denon, of course, with whom Forbin seems to have had a genuinely friendly relationship,[4] and Chateaubriand who, a little earlier, had published his *Itinéraire de Paris à Jérusalem*.[5] Chateaubriand and Forbin had met in Geneva in 1805, just before the writer's departure to the East.[6]

Forbin's journey followed the same route as the *Itinéraire*: Greece, the Archipelago, Asia Minor, Palestine and Egypt. Like Chateaubriand, his journey was a personal initiative, but he was officially commissioned to paint and bring back antiquities, a mission that allowed him to travel on a royal frigate of the Levant squadron and to receive important funds. The expedition was very well staffed. There was, for example, a panorama painter, Pierre Prévost, and an architect, Jean-Nicolas Huyot, each with an assistant. On board was also a young midshipman cadet, L.M.A. Linant de Bellefonds, who did not yet know that this journey would be a turning point in his life.[7]

The frigate *La Cléopâtre* left Toulon on 22 August 1817 for Greece. On 2 September she cast anchor in the roads of the island of Milo. Forbin took time to draw the ruins of the small ancient theatre; he also bought two fragments of female statues. While drawing these ruins, he could not have known that three years later, a few feet away, a Greek peasant farmer would discover another female statue that was to become one of the jewels of the Louvre, the famous 'Vénus de Milo'. Coincidentally, it was Forbin's son-in-law-to-be, the comte de Marcellus,[8] who finally managed to buy the statue and send it to France, in the midst of 'complex intrigues'.[9]

On 4 September, the travellers left Milo without Huyot, who had badly injured a leg, and sailed directly for Smyrna. They reached Athens on the 6th. Forbin was greeted by Louis Fauvel who, a few years earlier, had been Chateaubriand's host.[10] A painter and friend of Choiseul-Gouffier, Fauvel had lived in Athens since 1784 and had held the post of French consul since 1803. He had carried out many excavations in Greece and held a large slice of the antiquities market.[11] His house, situated in the centre of the ancient agora, was a real museum.[12] Guided by Fauvel, Forbin undertook a full visit of the ruins of Athens and even attempted a few excavations at the Piraeus, without much success. By the beginning of that century, Athens had become a fashionable meeting place for tourists of all nationalities. This irritated Forbin and gave him the opportunity to make a barbed remark about his arch-rivals, the British:

> Je rencontrai à Athènes des Anglais riches, dont l'affaire importante était de traverser la Grèce le plus promptement possible. J'y trouvai aussi plusieurs artistes anglais ou allemands dessinant, mesurant, depuis plusieurs années, avec l'exactitude minutieuse des commentateurs les plus scrupuleux, ces monuments, nobles créations du génie. Esclaves malheureux des règles, des moindres caprices des anciens, ils écrivent des volumes pour relever une erreur de trios lignes... sur la mesure d'une architrave; ils s'appesantissent, s'endorment et demeurent huit ans à Athènes pour dessiner trois colonnes.[13]

It must be said that Forbin himself did not hang around: he sailed again on 23 September, though not before buying part of the Fauvel collection for the Louvre, including a series of quite remarkable funerary reliefs.[14]

On 28 September Forbin arrived in Constantinople, where he was the guest of the Marquis de Rivière, the French ambassador to the Porte, who was also to play a significant role in the Venus de Milo affair. It was indeed on his behalf that Marcellus negotiated the statue. Constantinople absolutely dazzled Forbin, who described its cosmopolitan and bewitching atmosphere in a few concise but highly evocative lines:

> J'ai vu dans cette ville singulière des palais d'une admirable élégance, des fontaines enchantées, des rues sales et étroites, des baraques hideuses et des arbres superbes... Partout le Turc me coudoyait, le Juif se prosternait devant moi, le Grec me souriait, l'Arménien

voulait me tromper, les chiens me poursuivaient, les tourterelles venaient avec confiance se poser sur mon épaule; partout enfin on dansait et on mourait autour de nous. J'ai entrevu les mosquées les plus célèbres, leurs parvis, leurs portiques de marbre soutenus par des forêts de colonnes et rafraîchis par des eaux jaillissantes. Quelques monuments mystérieux, restes de la ville de Constantin, noircis, rougis par les incendies, sont cachés dans des maisons peintes, bariolées et souvent à demi brûlées. Les figures, les costumes, les usages, offrent partout le spectacle le plus pittoresque, le plus varié. C'est Tyr, c'est Bagdad, c'est le grand marché de l'Orient.'[15]

Fearing the plague, Forbin hastened his departure from Constantinople, leaving on 15 October. On the 20th he landed in Smyrna, where he met up again with Huyot who, though very well cared for by the missionary monks of the town, remained quite weak and as yet unable to continue the journey. The main reason for this stopover was to visit the ruins of Ephesus. Though Forbin describes them at length, his book only contains one drawing, that of the 'Persecution Gate', on the Ayasoluk hill (Fig. 8). This gate, which probably dates back to Justinian, was built with stones borrowed from ancient monuments: 'deux bas-reliefs', Forbin writes, 'étaient placés sans régularité au-dessus de la porte; les Anglais ont emporté depuis peu celui qui représente la mort d'Hector: l'opération a été si maladroitement dirigée, que le char d'Achille et le corps d'Hector restaient encore à prendre.'[16] In his turn, Forbin attempted the operation: something the English would not have done! But without ropes and tools, he was forced to give up. We now know that these reliefs came from an Attic sarcophagus that had been dismantled and set into the 'Gate of Persecution' during its construction. They reached England via Smyrna and Malta and were bought by the Duke of Bedford between 1822 and 1828 for his estate at Woburn Abbey, where they remain to this day.[17]

Forbin would have liked to extend his stay in Asia Minor and visit other ancient sites such as Sardes and Magnesia. In fact, Huyot's injury affected him greatly and he admitted that, had the architect been able to fulfil his commitments, the journey would probably have been more successful from an archaeological viewpoint. So he sailed again on 29 October, this time towards Palestine. On 6 November, after sailing past the islands of Chio and Rhodes, he landed in Saint Jean d'Acre and from there went by caravan to Caesarea, then on to Jaffa. By the middle of the month he had arrived in Jerusalem, where he spent a couple of weeks drawing and exploring the town, and also going on a few trips to Bethlehem, Jericho, the River Jordan and the Dead Sea. He found his visit to the Holy Sepulchre especially moving and produced several drawings: 'Il est impossible de n'être pas profondément ému', he wrote, 'de n'être pas saisi d'un respect religieux, à la vue de cet humble tombeau, dont la possession a été plus disputée que celle des plus beaux trônes de la terre.'[18] He also painted a general view of the town seen from the Josaphat valley [sic] (Fig. 9), of which Chateaubriand later said: 'En véritable peintre, M. le comte de Forbin a saisi le moment d'un orage, et c'est à la lueur de la foudre qu'il nous montre la cite des miracles.'[19] But as a writer, Forbin does not feel equal to the task: 'Je n'essaierai point de peindre Jérusalem après le grand écrivain dont la plume brillante et animée en a fait un si admirable tableau.'[20]

Forbin left Jerusalem on 2 December. Once in Jaffa, he decided to go to Damiette overland, which he achieved with great difficulties. The caravan moved off towards Gaza, but Forbin broke away from it just long enough to visit the ruins of Ascalon, which reminded him of historical events such as the presence of the Crusaders, Bonaparte's stay and Lady Stanhope's excavations. From the site he brought back the subject for a picture that he painted a few years later and which is now in the Louvre.²¹ The next stop was Gaza where, Forbin related, a 300-man retinue was being assembled around an English traveller, William John Bankes, who was about to visit the Nabateans' country. It is difficult to tell whether this inflated figure came from Forbin's informants or from Forbin himself, who was almost pathologically jealous and did not miss a single opportunity to point out the generous resources enjoyed by his English rivals. On 9 December, he left Gaza and got his first taste of a journey on camel back. But the hardest was yet to come: after a six-day walk, the caravan guides became lost in a maze of canals and marshes devastated by a recent hurricane. The young Linant, who was a trained sailor, swam off to get help. Finally, our travellers used a modest fishing boat to reach the Egyptian shore at Damiette, around 15 December. Forbin was shocked by the town's poverty:

> Les rues sont étroites et sans pavé, les maisons construites en brique, et toutes à demi détruites. Il est impossible de marcher sans craindre la chute de quelque corps avancé, de quelque poutre vermoulue: tout est en poussière ou en pourriture; les mosquées n'ont plus de portes, et les minarets menacent d'écraser des voûtes déjà entr'ouvertes.²²

In contrast, the house of the French consul, the rich Syrian merchant Basile Fackr, who received Forbin, looked like a palace.

On 22 December, our travellers left Damiette to sail up the Nile aboard a *djerme* and on Boxing Day they landed in Boulaq. There they found the vanguard of the Mecca caravan: the painter's eyes were seduced by the dazzling spectacle of the crowds with Cairo in the background. Feeling safe in his Muslim costume, Forbin did not hesitate to defy a few taboos: he visited all the Islamic monuments of the capital in great detail and spent hours in the Turkish baths, about which he left some rather contradictory accounts. At first he appears to have been totally seduced by them, the artist in him being especially sensitive to the architecture of the domes and the graceful patterns of the mosaics; but then he exclaims with disgust: 'Les bains publics sont spécialement le théâtre de ces débauches hideuses.'[23] This denunciation, which was shared by some of his Western contemporaries such as Pascal Coste,[24] was much influenced by the prejudices of the times. Forbin also ventured into the slave market, passing himself off as an Osmanli or a northern Turk, and even wondered about buying a pretty Circassian girl for 6,000 piastres. When he wanted a bit of solitude, he would leave the town to go and draw the caliphs' tombs (Fig. 10), which at the time were totally deserted: 'personne', he wrote, 'ne vient prier sur ces sépulcres de jaspe, au milieu de ces ruines vastes, silencieuses et presque inconnues aujourd'hui.'[25] Of course, Forbin also went to Giza, where the sight of the pyramids inspired him to write a few lyrical pages. But in front of the Sphinx, he couldn't help throwing another barbed remark at the British:

> J'arrivai trop tard pour profiter des travaux de M. Salt. Le déblaiement de la base de cette statue lui avait fait trouver un escalier qui communiquait à la porte d'un petit temple placé entre les pieds du sphinx; l'égoïsme le moins excusable venait de faire cacher de nouveau ce qu'il aurait été si curieux d'étudier, ce qui aurait jeté un si grand jour sur l'un des plus beaux monuments de la puissance des arts dans l'ancienne Egypte.[26]

After a couple of weeks in Cairo, Forbin sailed on 13 January for Upper Egypt: he hired a *cangia* in the company of a young Tuscan doctor, Martini, and the second dragoman of the French consulate, Ismail Rechouan, who had already been a guide for Chateaubriand. This Arabic name actually concealed a former drummer in the French army, Pierre Gary, who stayed on in Egypt after the Napoleonic troops departed. The travellers sailed up to Assiout where Forbin decided to continue his journey overland. On 28 January he reached Luxor, his ultimate destination.

At the beginning of that year, 1818, Luxor was a veritable archaeological battlefield. On one side was the English team, led by consul Salt, and on the other its French rival, led by Drovetti. Forbin noted, with undisguised jealousy, that the English were solidly ensconced in this area:

> Beaucoup d'argent, beaucoup de présents, leur ont conquis l'affection des Arabes, et toutes les entreprises ont été jusqu'à ce jour couronnées d'un succès complet. M. Drovetti lutte cependant avec persévérance contre ces nouveaux maîtres de l'Egypte. Il avait deux agents à Thèbes: l'un Youssef, mamlouk français, exploitait la rive occidentale du Nil; l'autre, nommé Riffo, fouillait l'enceinte du temple de Karnak.[27]

Forbin made a drawing of the gate of the great temple of Amon (Fig. 11): it was near here that Rifaud's excavations were so fruitful. Forbin describes Rifaud with amusement and slight exaggeration as a courageous, enterprising and quick-tempered little man, who, he says, would beat up the Arabs who persisted in not understanding the Provençal language.[28] He had 200 men working for him; Forbin hired 100 of them for a few days' excavations but admitted himself that he had no more success than at the Piraeus. Forbin also met Joseph Rosignani, known as Youssef, the 'French Mameluke', an ancient drummer in the Egyptian army, who lived in a cave in Gournah in the company of a young slave from the Ababdeh tribe.

Forbin made a point of securing a sketch of the French

team for posterity (Fig. 12). Wishing to be included in this drawing, he entrusted the task to the artist Jean-Pierre Granger. A lot of ink has been spilled over the identification of the people shown in this drawing. The only easily recognisable ones are Drovetti, dressed in European clothes and holding the plumb line, and behind him Rifaud, with an energetic face, a moustache and large sideburns. The other two Europeans on the left, both wearing Turkish costume, are probably Rosignani and the Piedmontese Antonio Lebolo, agents of the French Consul. The man leaning against the colossal head, on the right, is undoubtedly Forbin himself, since we know that he was travelling in an oriental costume: he ensured he was represented as a lord, absent-mindedly giving alms to a beggar woman. Standing near him is a Nubian, with his characteristic mass of hair, and, in the background, a young European in oriental costume. Some have seen Linant de Bellefonds in this young man, but this is impossible, since he remained in Cairo to help Prévost in painting the panorama of the town and thus did not accompany Forbin to Upper Egypt. I would rather suggest Frédéric Cailliaud, whom Forbin met in Luxor on Cailliaud's return from an expedition in the Arabian desert, where he had been sent by Muhammad Ali to search for emerald mines.

Forbin does not say how long he remained in Luxor, but he probably stayed no more than a fortnight, which he spent scouring the ruins, pencil and pad in hand. He would happily have stayed longer, and was even planning, he said, to sail up the Nile to Abu Simbel and, (why not?), to go as far as Meroe, but he gave up on the idea, saying that he was profoundly unsettled by the presence of English travellers and tourists – primarily the Belmores and their large entourage – who were returning from the cataracts. From then on, the expedition lost all its adventurous charm by becoming more accessible; and Forbin describes the haunting vision, as in a nightmare, of an English waiting-woman in a rose-coloured spencer,

Figure 7: *The comte de Forbin, by Paulin Guérin.*

Figure 8: *Ephesus, the Persecution Gate (after* Voyage dans le Levant, *pl. 6)*.

Figure 9: *General view of Jerusalem (after* Voyage dans le Levant, *pl. 17)*.

Figure 10: *Cairo, the caliphs' tombs (after* Voyage dans le Levant, *pl. 49).*

Figure 11: *Karnak, the great temple of Amon (after* Voyage dans le Levant, *pl. 60).*

Figure 12: *Drovetti, Forbin and the French team in Thebes, by Granger (after* Voyage dans le Levant, *pl. 73).*

Figure 13: *Alexandria, Cleopatra's Needles (after* Voyage dans le Levant, *pl. 76).*

Figure 14: *Muhammad Ali, drawing by Forbin (after G. Wiet,* Mohammed Ali et les Beaux-Arts *[Le Caire] pl. XLII).*

Figure 15: *The Mamelukes' massacre*, by Horace Vernet *(after* Voyage dans le Levant, *pl. 55).*

walking through the ruins holding a parasol.[29] That same evening, he sailed for Cairo.

This pretext may appear fallacious, and it certainly was, since we know, from a letter Forbin wrote to Drovetti on 27 December, that he did not intend to spend more than 40 days in Egypt.[30] The real reason for his hasty departure, in addition to sheer weariness, is probably that Forbin was very disappointed by the meagre results of his excavations in Luxor. In any case, Belzoni did not fail to make fun of him:

> Il assure gravement qu'il a été dégoûté de visiter les ruines de Louxor, en voyant s'y promener des Anglaises en spencer et avec des parasols. Voilà une plaisante raison pour un savant voyageur! Quel amour peut avoir pour les arts un homme qui s'enfuit à la vue de quelques Européennes, et s'excuse à son retour en Europe par ce motif bizarre de n'avoir pas pénétré plus en avant dans l'Afrique ?[31]

Belzoni probably had good reason to bear a grudge against Forbin. The count arrived in Cairo at the same time that Belzoni first tried, in vain, to open the pyramid of Kephren. In a sarcastic tone of voice, according to Belzoni, Forbin asked him to draw up a plan of the monument when he managed to enter it. A few days later, this feat accomplished, Belzoni sent the plan of the pyramid to Forbin, who was then in Alexandria, waiting for the ship that was to take him back to France. Forbin published the plan in his book as an *avant-première*, arousing the wrath of Belzoni, who accused him of having tried to claim credit for the discovery. To be fair to Forbin, it should be pointed out that he took care to caption the drawing in a way that left no doubt as to the ownership of the discovery, and that he added a laudatory comment: 'Le plan de l'ouverture de la seconde pyramide de Gyzeh donne une idée exacte des efforts et de la marche de M. Belzoni dans cette découverte importante... Ce jeune Romain a six pieds; sa force physique est extraordinaire: il est intelligent, entreprenant, courageux, et très-dévoué au gouvernement britannique.'[32] This last quality was probably, in Forbin's eyes, Belzoni's main failing!

After a few days in Cairo, Forbin went to Alexandria, his first visit to the city. While drawing 'Cleopatra's Needles' (Fig. 13), the two obelisks that marked the entrance to the Caesareum, he noted with undisguised pleasure: 'Les Anglais viennent de faire de vains efforts pour remuer l'obélisque couché, dont on aperçoit la base à côté de celui qui demeure encore debout.'[33] However, whatever Forbin's views on the matter, this respite did not last. Eventually the obelisk came to rest on the banks of the Thames, while its standing companion was offered to the United States by the Khedive Ismail.

Before leaving for France, Forbin studied carefully the collection of antiquities assembled by Drovetti: 'Sa collection d'antiquités égyptiennes est admirable', he wrote, 'et son voeu le plus vif serait d'embellir le musée de Paris; c'est dans cette espérance qu'il a sans cesse refusé de la vendre, malgré les offres brillantes qui lui ont été faites.'[34] As soon as he was back in Paris, Forbin tried many times to persuade his government to buy the collection. These attempts were always thwarted by Louis XVIII's economy drive and by the advocates of Greek classicism, who saw no artistic qualities in Egyptian monuments. The Drovetti collection was eventually acquired by the king of Sardinia and exhibited in Turin.

However, Forbin did not return from Egypt empty-handed. In Cairo, Belzoni, who badly needed money to continue his work, agreed to sell him two of the splendid Sekhmets he had discovered during his excavation of the Mut temple in Karnak. Belzoni said of this sale that the price paid by Forbin was not a quarter of the value of these antiques, but as he had never sold any statues, he was satisfied with the deal.[35] Riffaud had also excavated the Mut temple, in a different area from Belzoni, and within a few weeks had unearthed four large, practically intact Sekhmets: two of these went to Alexandria to join the Drovetti collection, while the other two remained in Cairo and were also acquired by Forbin for the Louvre. Finally, in Gournah Forbin had bought two mummies from Joseph Rosignani. All these acquisitions were transported by Drovetti to Alexandria, and then loaded on a ship for France.[36]

In addition to the antiquities acquired for the Louvre in Greece and in Egypt, the main interest of his journey lies in the luxurious book Forbin published barely one year after his return to France. He surrounded himself with an impressive group of artists – painters, aquarellists and lithographers – to produce the 80 plates selected from his portfolio of drawings. The text, as he underlines with a touch of modesty, should only be seen as a commentary to accompany the illustrations. Admittedly, Forbin was not a professional writer, but his lively and pleasant account is that of an enlightened observer. He compares himself to a journalist, thus pre-empting the criticisms of those who reproached him for having produced a superficial work[37]: 'Les voyageurs sont, pour ainsi dire, les journalistes du monde... j'écris ce que je vois, ce qui me frappe. Ces notes, si insuffisantes pour les autres, ne sont que le miroir de mes impressions.'[38]

While archaeological finds were the official purpose of Forbin's journey, the Orient of the times often takes priority in his story which, in spite of his hasty and often exaggerated judgements, remains a precious account of the Egypt of Muhammad Ali. It would be true to say that Forbin helped to disseminate, throughout the Western world, a certain image of the Pasha, whom he did not hesitate to compare to a new pharaoh: 'Les Egyptiens sont encore ce qu'ils étaient sous le sceptre des Pharaons; c'est pour un maître qu'ils cultivent leurs terres, c'est pour lui qu'ils couvrent le Nil de bateaux. Le fellâh sert aujourd'hui la cupidité de Mohamed Aly, comme il obéissait jadis à l'orgueilleuse volonté qui faisait construire les pyramides.'[39]

Before leaving Egypt, and thanks to Drovetti, Forbin obtained several audiences with the Pasha; he even obtained the favour of painting his portrait (Fig. 14). It is interesting to note that Forbin did not choose to have a lithograph made of this drawing for his book.[40] But this same portrait was used as a basis for the famous painting of the Mamelukes' massacre, commissioned by Forbin from his friend, the painter Horace Vernet (Fig. 15). Reproduced in his book and much commented on later, this work popularised in Europe the image of Muhammad Ali as the oriental despot placidly watching the massacre ordered by him. As a pure artistic reconstruction of a historical event, it is a perfect metaphor for Forbin's approach: his journey was not an archaeologist's, nor an adventurer's, nor a journalist's. It was, above all, that of a painter. Chateaubriand stressed: 'M. le comte de Forbin, dans son *Voyage au Levant*, réunit le double mérite du peintre et de l'écrivain... Notre course, quasi la même, a été accomplie dans le même espace de temps. Plus heureux que nous seulement, M. le comte de Forbin avait un pinceau pour peindre; et nous, nous n'avions qu'un crayon.'[41]

# REFERENCES

1. *Quarterly Review,* May 1820, p. 83.
2. Very tall, with an elegant and noble demeanour, regular features reminiscent of the handsome faces of the century of Louis XV, he was what you would have called, in the old days, an accomplished gentleman.' Siméon, 'Notice historique sur M. le comte de Forbin, lue à l'Académie des Beaux-Arts le 27 mars 1841', *Moniteur universel,* 28 March 1841.
3. 'Incredible circumstances had erected a huge monument. Similarly extraordinary circumstances have just toppled it. Europe had to be defeated to constitute this trophy; then Europe reunited to destroy it.' Quoted from Ph. Sollers, *Le Cavalier du Louvre. Vivant Denon* (Paris, Plon, 1995), p. 247.
4. Denon appreciated Forbin's talent, since he owned one painting and three drawings by him: A. N. Pérignon, *Description des objets d'art qui composent le cabinet de Feu M. le baron V. Denon* (Paris, printed by H. Tillard, 1826), no. 201 (painting), 861 (two drawings), 862 (drawing).
5. F. de Chateaubriand, *Itinéraire de Paris à Jérusalem et de Jérusalem à Paris, en allant par la Grèce, et revenant par l'Egypte, la Barbarie et l'Espagne* (Paris: Le Normant, 1811).
6. 'M. de Forbin était alors dans la béatitude; il promenait dans ses regards le bonheur intérieur qui l'inondait; il ne touchait pas terre. Porté par ses talents et ses félicités, il descendait de la montagne comme du ciel, veste de peintre en justaucorps, palette au pouce, pinceaux en carquois. Bonhomme néanmoins, quoique excessivement heureux, se préparant à m'imiter un jour, quand j'aurais fait le voyage de Syrie, voulant même aller jusqu'à Calcutta, pour faire revenir les amours par une route extraordinaire, lorsqu'ils manqueraient dans les sentiers battus.' F. de Chateaubriand, *Mémoires d'Outre-Tombe* ed. M. Levaillant, Paris: Flammarion, 1949), I 2, p. 195.
7. M. Kurz and P. Linant de Bellefonds, 'Linant de Bellefonds: Travels in Egypt, Sudan and Arabia Petraea, 1818-1828', in P. and Starkey (eds.), *Travellers in Egypt* (London: I.B. Tauris, 1998), pp. 61-69.
8. Lodoïs de Martin du Tyrac, comte de Marcellus, married Valentine de Forbin in 1824.

9. A. Pasquier, *La Vénus de Milo et les Aphrodites du Louvre* (Paris, Réunion des musées nationaux, 1985), pp. 24-28; C. Doumet-Serhal, 'Rencontre à Sidon avec Lady Esther Stanhope. Le comte de Marcellus et la Vénus de Milo en transit', *Archaeology and History in Lebanon* 12, 2000, pp. 24-34.
10. A. Outrey, 'Note sur la maison habitée par Chateaubriand pendant son séjour à Athènes au mois d'août 1806', *Bulletin de la Société Chateaubriand*, 25 juin 1936, pp. 1-6.
11. P.E. Legrand, 'Biographie de Louis-François-Sébastien Fauvel, antiquaire et consul (1753-1838)', *Revue archéologique* 30, janvier-juin 1897, pp. 41-67, 185-201, 385-404; 31, juillet-décembre 1897, pp. 94-103, 185-223.
12. See the lithograph entitled 'L'Acropole, vue de la maison du consul de France, M. Fauvel, 1819' in Louis Dupré, *Voyage à Athènes et Constantinople* (Paris: printed by Dondey-Dupré, 1825), pl. 19.
13. 'In Athens I met rich English people whose main objective was to travel across Greece as quickly as possible. I also came across some English and German artists who had spent several years drawing and measuring these monuments, these noble creations of genius, with the meticulous accuracy of the most scrupulous commentators. Pathetic slaves to the rules, to the smallest whims of the ancients, they write volumes to point out a three-line mistake in the dimensions of an architrave; they dwell on details, become sleepy and stay eight years in Athens to draw three columns.' Comte de Forbin, *Voyage dans le Levant en 1817 et 1818* (Paris: Imprimerie royale, 1819), p. 35.
14. M. Hamiaux, *Les sculptures grecques du Louvre I* (Paris, Réunion des musées nationaux, 2002), No. 140, 144, 146, 183, 199.
15. 'In this singular city I saw palaces of admirable elegance, enchanted fountains, dirty and narrow streets, hideous houses and beautiful trees... Everywhere I rubbed shoulders with Turks, Jews prostrated before me, Greeks smiled at me, Armenians tried to deceive me, dogs followed me, turtle doves came to rest on my shoulder with complete trust, and everywhere people danced and died around us. I caught glimpse of the most famous mosques, their courtyards, their marble porticos supported by forests of columns and cooled by gushing

waters. A few mysterious monuments, the remnants of Constantine's city, blackened and reddened by fires, are hidden behind gaudily painted and often half-burned houses. Everywhere the faces, the costumes, the customs offer the most picturesque, the most varied spectacle. It is Tyre, it is Baghdad, it is the great Eastern market.' *Voyage dans le Levant*, p. 44.

16. 'Two reliefs were positioned irregurlarly above the gate; the English recently removed the one representing Hector's death: the operation was so badly organised that Achilles' chariot and Hector's body are still there.' *Voyage dans le Levant*, p. 61.
17. Although the identity of the individual who removed the reliefs remains uncertain, it might have been the Reverend F.V.J. Arundell, the British chaplain at Smyrna: E. Angelicoussis, *The Woburn Abbey Collection of Classical Antiquities* (Mainz am Rhein: Verlag Philipp von Zabern, 1992), pp. 79-80; on other mentions of the 'Persecution Gate' by French travellers: R. Chevallier, 'La porte de la Persécution à Ephèse: à propos d'une vue du dessinateur L.F. Cassas', in H. Friesinger and F. Krinzinger, ed., *100 Jahre Österreichische Forschungen in Ephesos* (Vienna, Verlag der Österreichischen Akademie der Wissenschaften, 1999), p. 655-660.
18. 'It is impossible not to be profoundly moved, not to be overwhelmed by religious respect at the sight of this humble tomb, the possession of which has been more disputed than that of the most beautiful thrones in the world.' *Voyage dans le Levant*, p. 106.
19. 'Like a true painter, M. the Comte de Forbin has caught a passing thunderstorm and manages to show us the city of miracles lit up by lightning.' F. de Chateaubriand, *Mélanges littéraires* (Œuvres complètes, Paris, Ladvocat, 1826-1831), vol. XXI, p. 382.
20. 'After the great writer whose brilliant and spirited pen has made of it such a wonderful picture, I will not attempt to depict Jerusalem.' *Voyage dans le Levant*, p. 84.
21. Inv. 4490. The picture is entitled: 'Ruines d'Ascalon en Syrie, effet de soleil levant'.
22. 'The streets are narrow and unpaved, the houses are made of bricks and are all partially destroyed. It is impossible to walk around without the fear of being hit by some falling object or worm-eaten beam: everything

is decaying or reduced to dust; mosques have no doors and their minarets look as if they are about to crash down on already broken vaults.' *Voyage dans le Levant*, pp. 192-193.
23. 'These baths are notable for being the theatre of hideous debauchery.' *Voyage dans le Levant*, p. 227.
24. Pascal Xavier Coste, *Architecture arabe ou monuments du Kaire, mesurés et dessinés, de 1818 à 1826* (Paris, Firmin-Didot, 1837), pp. 42-43, 'Bains publics, planche LIII': 'Plus que dans aucun pays, les bains sont au Caire un lieu de débauche pour les hommes.'
25. 'Nobody comes here to pray on these jasper sepulchres, in the middle of these vast, silent and now almost unknow ruins.' *Voyage dans le Levant*, p. 228.
26. 'I arrived too late to benefit from the labours of Mr Salt. Clearing the base of this statue, he found a staircase leading to the door of a small temple situated between the legs of the sphinx; the most unpardonable egotism made him hide again something that would have been so interesting to study.' *Voyage dans le Levant*, p. 218. For Caviglia's reaction when this accusation reached his ears, see: D. Manley and P. Rée, *Henry Salt* (London: Libri, 2001), p. 108.
27. 'A number of presents and yet a more profuse distribution of money have overpowered the barren affection of the Arabs, and all their enterprises among them have succeeded in a wonderful manner. Meanwhile, Mr Drovetti is fighting these new masters of Egypt with persevering consistency. He had two agents in Thebes: one, Youssef, a French mameluke, was exploiting the west bank of the Nile; the other, named Riffo, was digging the wall of Karnak temple.' *Voyage dans le Levant*, pp. 266-267.
28. *Voyage dans le Levant*, p. 267.
29. *Voyage dans le Levant*, p. 273.
30. B. Drovetti, *Epistolario*, ed. S. Curto and L. Donatelli (Milano, Cisalpino-Goliardica, 1985), no. 73. When Salt saw Forbin's book, he immediately wrote to Lord Belmore: D. Manley and P. Rée, *Henry Salt*, p. 139.
31. 'He gravely tells us that he was disgusted, while visiting the ruins of Luxor, at the sight of English women wearing a spencer and holding a parasol. This is a strange reason for an erudite traveller! How can a

man who has crossed the seas to see the wonders of ancient Egypt genuinely love the arts if he runs away at the sight of a few European women and, back in Europe, gives this as a reason for not going any further into Africa?' G. Belzoni, *Voyages en Egypte et en Nubie* (Paris, G.P. Depping, 1821, reed. Paris: Pygmalion, 1979), p. 202.

32. 'The plan of the opening of the second pyramid of Giza gives an accurate idea of the efforts and the role of Mr Belzoni in this important discovery. This young Roman is six foot tall; his physical strength is extraordinary: he is intelligent, enterprising, courageous and fully devoted to the British government.' *Voyage dans le Levant*, pp. 459-460.
33. 'The British have just attempted, in vain, to move the fallen obelisk, the base of which we can see next to the one that is still standing.' *Voyage dans le Levant*, pp. 457-458.
34. 'His collection of Egyptian antiquities is admirable, and his greatest wish would be for it to enhance the Paris museum; it is in this hope that he always refused to sell it, despite the brilliant offers that have been made to him.' *Voyage dans le Levant*, p. 226.
35. Belzoni, *Voyages en Egypte et en Nubie*, op. cit., p. 201.
36. Apart from these objects acquired for the Louvre, Forbin may also have bought some antiquities for himself: J.-C. Goyon, 'Une curieuse histoire de cercueil', *Kyphi* 3, 2001, pp. 19-56.
37. Salt's criticisms were specially caustic: D. Manley and P. Rée, *Henry Salt*, p. 139 and n. 32.
38. 'Travellers could be said to be the journalists of the world... I write about what I see, what strikes me. These notes, so inadequate for some, simply convey my impressions.' *Voyage dans le Levant*, p. 248.
39. 'Egyptians are still what they were under the sceptre of the Pharaohs; they cultivate their land and cover the Nile with boats for a master. Nowadays, the fellah serves Muhamed Ali's cupidity just as, in days gone by, he obeyed the overproud will that led to the building of the pyramids.' *Voyage dans le Levant*, p. 248.
40. The portrait is reproduced in the *Bibliothèque Egyptologique*, vol. 31, p. 35, with the mention: 'Lithographie de C. Motte d'après un dessin de Mr le Comte de Forbin fait à Alexandrie en mars 1818.'

41. 'M the Comte de Forbin, in his *Travel to the Levant*, combines the double merit of being a painter and a writer... Our own journey, almost the same, was accomplished within the same time period. However, happily for him, Mr Forbin had a brush to paint with, while we only had a pencil.' Chateaubriand gave an account of Forbin's book in *Le Conservateur*, 3, 1819, pp. 385-396.

# Chapter Five
# Travellers, Tribesmen and Troubles: Journeys to Petra, 1812-1914[1]

## Norman N. Lewis

The first European in modern times to reach Petra was J.L. Burckhardt.[2] In 1812, having spent three years in Syria improving his Arabic and habituating himself to dressing and behaving as an Arab, he set out from Damascus to travel to Cairo. He chose to follow a route east of the Dead Sea and through Kerak. It was probably in or near Kerak that a group of local people spoke to him 'in terms of great admiration' about the 'antiquities' at a place called Wadi Mousa. He decided to go and see them, and once there he realised that the ruins were almost certainly those of the ancient Petra.

Burckhardt met W.J. Bankes in Cairo in 1816 and told him about his Syrian journeys. Bankes was particularly intrigued by what Burckhardt told him about the ruins in Wadi Mousa and determined to go there. He was not able to do so until 1818, but then he, with Irby, Mangles and Legh, left Jerusalem on May 6 and reached Petra on the 24th. They were only able to do this because they had the extraordinary good fortune to obtain the support of Yousif Majalli, 'the lord of Kerak' (who had ill treated Burckhardt six years earlier). Muqabil Abou Zaitoun, 'the sheikh of Wadi Mousa', stubbornly opposed the idea that these strangers should enter his valley, but after several days of negotiation Sheikh Yousif and others managed to persuade him to allow the travellers to spend two days among the ruins.[3] In that short period Bankes made drawings of the Khazneh and other monuments, plus copies of several inscriptions, the first such ever made.

Bankes returned to Egypt later in 1818 and there got to know a young French artist, L.M. Linant de Bellefonds, with whom he doubtless discussed his Petra journey. Linant remained in Egypt (or travelling in neighbouring areas) for most of his life. In 1827 he met Léon de Laborde with whom, the following year, he went from Cairo via Sinai and Aqaba to Wadi Mousa. Once arrived in Wadi Mousa they found that the plague was raging and the local people left them alone. They were able to stay for eight days, exploring the site undisturbed and at leisure, and producing a map and numerous drawings.[4]

By travelling through Sinai and Aqaba to Petra, Laborde and Linant had pioneered a route which, with modifications, was eventually to be followed by the great majority of European visitors to Petra. Many tourists met Linant in Cairo in the 1830s and sought his advice. They already knew, no doubt, that the way to Mount Sinai and the Monastery of St. Katherine was well trodden and safe and Linant was able to tell them that they could go on from there across the Sinai peninsula to Aqaba where they could find guides, camels and escorts to take them to Petra. Amongst those he advised, or to whom he gave letters, were J.L. Stephens who reached Petra in March 1836 (he was the first American to go there)[5], the Rowley party (Charlotte Rowley was almost certainly the first woman visitor) later in the same year,[6] Lord Lindsay in the following year,[7] and

David Roberts the artist and his companion J.G. Kinnear who travelled in 1839.[8]

Until the end of the 1830s, most visitors to Petra travelled individually or in small parties but in the following decades the route became very popular and was used by large parties of tourists. An extended itinerary was now commonly followed: from Cairo to St. Katherine's Monastery, thence to Aqaba and Petra and finally to Hebron – 'the gateway to the Holy Land'. This journey could be performed in perfect accord with the seasons: after perhaps several weeks enjoying the benign winter weather of Egypt, visitors could begin their desert journeying early in the New Year. After four days or so getting to Suez (this was reduced to as many hours in the 1860s when the Cairo-Suez railway opened) they would spend at least a fortnight in Sinai, of which several days would be devoted to St. Katherine's and its vicinity. When they had crossed the peninsula and reached Aqaba they would probably find it necessary to stay there for a few days before setting out for Petra, some four days' journey to the north. By far the most popular time for visiting Petra was in March or April before the weather became too hot; over 50 per cent of the published accounts describe visits made in these months. Most people stayed there for two or three days, few for longer. Hebron was only four days away from Petra and Bethlehem and Jerusalem only a day or two beyond that. Spring weather in Jerusalem was usually agreeable and travellers who had planned their journey carefully could be in Jerusalem at Easter.

All or nearly all the journeys were done on camel back The indispensable preliminary, to be undertaken in Cairo or Aqaba, was to come to an agreement with a 'camel-sheikh' who would provide riding camels, baggage camels, cameleers and escorts. The sheikh or one of his deputies would lead the caravan. The first part of the journey – from Cairo to Suez, St. Katherine's Monastery and Aqaba –

could easily be arranged in Cairo with sheikhs of certain tribes of the Towara (the bedouin of Sinai) whose families had specialised for generations in providing transport and services to pilgrims making for St Katherine's, and they did their work well. Several of the travellers of the 1830s engaged the same sheikhs who had served Linant and Laborde, and were well satisfied. Throughout the century travellers continued to record the kindness and reliability of the Towara. The Towara could not, however, go beyond the limits of their own territory and once they had delivered their charges to Aqaba they had to leave them. From then on, things could become difficult and unpleasant for the travellers. Aqaba in those days was a miserable place consisting of a dilapidated fortress and a few hovels. The travellers would almost certainly have to camp there for several days, first until the next sheikh with whom they had to negotiate made his appearance, then while they tried to reach agreement with him on terms and finally while their caravan was made ready.

The sheikh with whom most of the travellers had to deal in Aqaba until his death in the early 1840s was Hussain Ibn Jad, sheikh of the Alouin, the bedouin tribe which dominated the eastern littoral of the gulf of Aqaba and much of the country between Aqaba and Petra. The Alouin were famed as caravaneers; each year they provided men and camels to the great Hajj caravan with which thousands of pilgrims travelled from Cairo to Aqaba and thence southwards to the Muslim holy cities. They were mentioned as early as 1738, by Pococke, who wrote that they were very bad people and notorious robbers. Later travellers (and bedouin of other tribes) described them similarly; in 1858 Porter called them 'as impudent and lawless a set of vagabonds as ever pilgrim had to deal with'.[9] Hussain himself was a skilful and ruthless bargainer (his conduct while negotiating with Formby and his party in 1840 was described as 'a perfect piece of cool, shrewd, sharp sighted, resolute cunning')[10] and the travellers were in a weak bargaining position. They were effectively marooned; short of attempting to return to Egypt or trying to reach southern Palestine, they had no choice but to go forward to Petra and they could only do that with Hussain's Alouin. A few resolute travellers took a firm line with him and managed to obtain a reasonable agreement but most of them ended up having to accept his terms even if they considered them outrageous. Hussain naturally insisted on taking an escort of armed men in case other bedouin attacked the caravan. Although such events were rare and some travellers believed that Hussain talked up the danger in order to extract more money from them, the risk was genuine; Hussain's own camp was raided by bedouin of another tribe in 1837 while he was escorting Lindsay and others on their way from Petra to Hebron.[11]

None of the travellers gives a complete account of the expenses incurred on the journey, and the figures that do appear are often imprecise and confusing. It appears, however, that the equivalent of ten English pounds for each riding camel and about a quarter of that for a baggage camel for the Aqaba-Petra-Hebron journey was commonly paid in the first half of the century; much larger figures are mentioned towards the end. A few people travelled light and more or less independently, but most joined groups which were accompanied by a dragoman (usually Egyptian), a cook and a number of servants of one kind or another, and they carried tents and other gear as well as vast amounts of food bought in Cairo and gradually consumed as they went. On top of the cost of all this came the fees and *bakhshish* which had to be paid at Aqaba, Hebron and particularly at Petra, where the nearest thing to an official entrance fee was the equivalent of an English pound, but where the actual sums paid were frequently much greater.

All the travellers were glad to get away from Aqaba and

**Figure 16:** *A typical group of Petra Bedouin c. 1895, from* Petra: Perea Phoenicia, *Rev A. Forder. Marshall Brothers, London and Edinburgh, book undated, but probably published about 1920, p. 15.*

**Figure 17** (left): *Petra guides, from* Petra: Perea Phoenicia: *Rev. A. Forder, p. 15.*

**Figure 18** (opposite): *Title page of Formby's* A Vist to the East, 1843, *with a sketch of Sheikh Muqabil abou Zaitoun 'Chief of the Inhabitants of Wadi Mousa'. Formby was at Petra in the spring of 1840; Muqabil died two or three years later.*

# A VISIT TO THE EAST;

COMPRISING

GERMANY AND THE DANUBE, CONSTANTINOPLE, ASIA MINOR, EGYPT, AND IDUMEA.

BY

## THE REV. HENRY FORMBY, M.A.

SHEICK YOMGEBEL ABOUSEETON,
CHIEF OF THE INHABITANTS OF WADI MOUSA.

## LONDON:
JAMES BURNS, 17 PORTMAN STREET.
M·DCCC·XLIII.

*Figure 19: Photograph of the Eastern Cliff at Petra taken by John Shaw Smith in 1852. Reproduced with the permission of the Palestine Exploration Fund.*

to be on the move again, but few of them enjoyed the first few days of the journey towards Petra. After the quiet, friendly Towara the Alouin came as a shock; the travellers described them variously as wild, fierce, lawless, disobedient, dirty, avaricious. They grumbled, quarrelled and shouted all day and, it seemed, all night. They scrounged and begged and bothered, and Sheikh Hussain or his representative sometimes seemed to be the most annoying of them all.

As the travellers approached their goal, feelings of apprehension – as well as those of pleasurable anticipation – made themselves felt. They knew that Petra was a dangerous place, beyond the reach of any government, and that as they went down into the valley they might be confronted by armed men, wilder looking than even the Alouin, barring the way and demanding '*ghufr*, that is a tax, tribute, present or whatever else it may be called, for the privilege of visiting the place'.[12] Some of the visitors did not object to paying what they thought of as an entrance fee – akin to paying to look over the Tower of London, as more than one American put it – so long as it was a reasonable amount. Others thought it 'an imposition' and balked at paying it. Then tempers might rise, and even if the visitors agreed to pay someone the 'entrance fee', that was seldom the end of it, for no one person or group of people was 'in charge' of the ruins. Often enough the visitors were assailed by a succession of claimants, all wanting a *ghufr* or *backhshish*, shouting and pushing and quarrelling amongst themselves.

A root cause of the difficulties lay in the tribal situation. Varying numbers of bedouin, most of them belonging to a very poor little tribe called the Bdoul or Bdoun, often camped amongst the ruins[13] ( see Figs 16 and 17). The fact that most of them were Bdoul did not mean that they obeyed one sheikh or were organised as an entity; there were four or more sub-tribes among them, and when it came to a scramble for a

**Figure 20:** *The group of artists and others led by Jean-Leon Gerome on his 'Expedition en Orient' in 1868. The picture was painted by W. de Famars Testas. Reproduced by permission of the Rijksmuseum van Oudheden, Leiden.*

# TRAVELLERS, TRIBESMEN AND TROUBLES

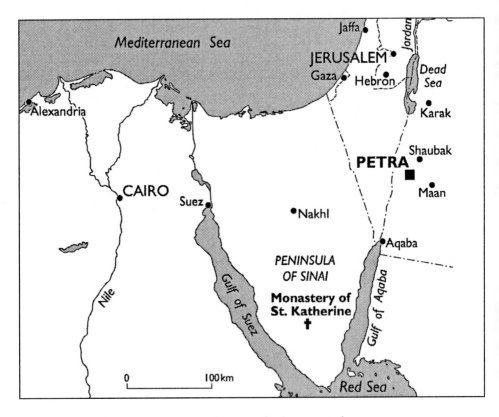

Figure 21: *Map showing the location of Petra.*

*ghufr* or *backhshish* it was every man for himself. A more numerous and formidable group of people, the Liathna, lived towards the upper end of the valley. They were farmers, not bedouin, although most of them lived in bedouin-style tents for much of the year. They grew grain and fruit on terraced, irrigated land around Alji, their main village, situated above the site of the present town of Wadi Mousa. The bedouin usually called them simply 'the *fellahin*'. They were a peculiar people, unrelated to other tribes, reputed to be cruel and malicious. Many of the Alouin and other bedouin were frankly afraid of them. In the first half of the 19th century their sheikh, Muqabil Abou Zaitoun, was very much a man to be reckoned with, as Bankes and his party had found in 1818 (Fig. 18). In dealing with later visitors he argued vehemently but cogently that Wadi Mousa, the ruins included, belonged to his people who lived there, and that they were entitled to benefit from the fact.[14] Unfortunately he seems to have been better at dealing with visitors than he was at controlling his own men, some of whom, like the bedouin, were inclined to run riot when parties of wealthy foreigners presented themselves for the plucking.

A final cause of trouble was, of course, mutual incomprehension. Few of the dragomans who accompanied tourists were effectively bilingual and only rare individuals among the Europeans spoke Arabic.

So it was that visitors would pay one man or group but then would be pestered by others. Quarrels and fights would break out between the different groups and the travellers might somehow get embroiled. Sometimes, if travellers had not paid enough when they arrived, trouble developed as they tried to leave and things could get quite out of hand as they did in 1858, as Edward Lear prepared to leave. Here are a few lines from his diary:

> They became more importunate and turbulent every minute, snatching at any object within reach and menacing

with their firearms... 20 or 30 of them dragged the camels away and then in another moment a larger number fell on the first party and were for a time masters... A large body of them rushed in a frenzy with deafening cries and hustled and dragged us and ... holding up my arms and unbuttoning my clothing extracted in a twinkling everything from all my many pockets from dollars and penknives to handkerchiefs and hardboiled eggs.[15]

One should perhaps take Lear's account with a pinch of salt but many visitors complained with good reason of the harassment which accompanied their arrival or departure and of the petty thefts and 'indignities' which they suffered. Others, however, had little or no trouble and enjoyed themselves thoroughly. Most of them hired local men to guide them about the ruins, and a few even visited the tents of the *fellahin* and were well received. John Shaw Smith, in 1852, was well content to ramble about quietly taking photographs (probably the first ever taken there and almost certainly the earliest which have survived, see Fig 19). Others preferred more robust activities such as shooting partridges and pigeons on the wing 'to the delight and astonishment of the Arabs'.[16]

The behaviour of the local people seems to have varied unpredictably and was probably affected by external factors unknown to the visitors, but it was also influenced by the behaviour of the visitors themselves, who differed greatly in their attitudes and reactions. Those of them who had good dragomans, who behaved sensibly and fairly and were reasonably open-handed, were more likely to get on well than those who were bellicose or penny pinching. The experiences of three parties in 1868 illustrate the point. In February the American artist Frederick Church spent a couple of days in Petra with his friend, Rev D.S. Dodge, who wrote that: 'The Arabs of Petra, though generally so annoying and even threatening to travellers, gave us no trouble whatever. Our experienced dragoman understood what cords to pull: "Behave well or no *bucksheesh*", and his firmness and liberality succeeded admirably.' Despite initial qualms Church found that he could 'sketch without let or hindrance' and that his bedouin guards were helpful to him.[17]

A month later two other parties arrived. The first was that of the Marchioness of Ely, a Lady-in-Waiting of Queen Victoria. She had no difficulty in ensuring that she enjoyed good treatment; she simply appropriated the then 'Sheikh of Petra' as her personal bodyguard – and rewarded him lavishly.[18] The second was that of the French painter, J.L. Gérome, with his 'caravan of artists' (Fig. 20) who arrived and left a day or two after Lady Ely. It rained a lot while they were there and they had to spend much of their time in camp where they found the noise and demands of the people (the 'vermin' or 'canaille') almost insufferable. Their relations with the sheikh deteriorated; he wanted more money than they were prepared to give (he was perhaps hoping for something approaching Lady Ely's extravagant generosity). When they tried to leave, he and his men barred the way and threatened them and they only got away after a dramatic confrontation with much flourishing of firearms on both sides. Paul Lenoir, one of the party, afterwards wrote that at the time their ambition was 'to return to Petra at some future day and exterminate these banditti to the last man.'[19]

Whether or not people had enjoyed themselves in Petra they were glad to start on their last lap of four days or so to Hebron. This usually seemed relatively easy to the now hardened travellers. They were buoyed up by the thought that soon there would be no more camel riding to be done and no more Alouin (though some of the travellers had after-thoughts – maybe the Alouin hadn't been all that bad and had after all got them safely to the end of the journey). The Alouin were paid off at Hebron, and a few days later the travellers would spruce themselves up and ride, on horseback at last, towards the Holy City of Jerusalem.

Not all visitors to Petra travelled from Aqaba with the Alouin; some independent travellers deliberately avoided the necessity of doing so by striking north from central Sinai to Nakhl, a little way-station on the Cairo-Aqaba pilgrim route, and thence eastward to Petra. Others found their way from the coast of southern Palestine. Journeys by these routes were difficult; travellers had to deal with several tribes en route. The journey from Jerusalem to Petra was usually a little easier, but once the travellers reached Petra they were liable to experience the same troubles as those who had come from Aqaba. In 1851 James Finn, the British Consul in Jerusalem, undertook the journey himself in the hope of extracting a promise from the then sheikh of the *fellahin* that his people would treat visitors from Jerusalem better in future.[20] He secured some kind of oral agreement from the sheikh but it made no apparent difference; Edward Lear, who came from Jerusalem in 1858 (to get away from the Easter crowds) was not the only such person to be badly treated in the next few years.

1868 was a busy year for the Sinai-Aqaba-Petra-Jerusalem route, but after that it was much less frequently used. With the opening of the Suez Canal in 1869 and the development of steamship communication, the tourist trade between Europe, Egypt and Palestine flourished. The steamship route from Port Said or Alexandria to Jaffa (whence the road to Jerusalem was rebuilt) now afforded a quick and easy way of getting to Jerusalem. Mount Sinai and St. Katherine's Monastery could now be approached from the port of Tur on the Gulf of Suez, or the old land route could be used for the southward journey and a ship for the return. (It was more important to the people of Egypt and other Muslim countries that they could now take ship to Jiddah in order to reach Mecca and Medina; few of them regretted that the Alouin were thus deprived of their main source of income).

The 1876 edition of Cook's *Tourists' Handbook for Egypt* reflected the changed conditions. It was unenthusiastic about the desert route from St. Katherine's Monastery to Jerusalem via Petra; this, it advised, 'should not be persisted in unless intelligence is received from 'Akabah that all is well. From Mount Sinai... it is far preferable to return to Suez, proceed by the canal to Port Said and then take the steamer to Jaffa'.[21] All was often not well at Petra or in the country round about it. In 1870 the *fellahin* of Wadi Mousa were said to be 'more lawless and turbulent than ever'. Two years later Canon Tristram was briefly 'detained and held to ransom' at Kerak, and in 1882 Palmer and two of his companions were murdered near Nakhl. In the 1870s and 1880s the countryside around Kerak and south of it was intermittently plagued by tribal 'warfare' in which the people of Wadi Mousa were sometimes involved. Few Europeans visited Petra in this period; Hull, Kitchener and other members of the Palestine Exploration Fund's scientific expedition of 1883 were notable exceptions.

In 1893 the Ottoman authorities decided to put an end to the strife and to impose their authority on the area east and south-east of the Dead Sea. In November Turkish troops occupied Kerak and an administrative centre was afterwards established there. Troops were also sent to Maan and Shaubak. In 1895 the people of Shaubak attempted a revolt, but the army bombarded their town and severely chastised the local bedouin (a second, more serious rebellion erupted in Kerak in 1910 and was put down). Once order had been imposed, Turkish officials in Kerak, Maan and Shaubak received occasional European visitors politely and gave them permits to visit Petra, together with a soldier or two to accompany them. The first such visitors appear to have been Forder and Hornstein, who reached Petra from Kerak in September 1895.[22] They were followed by Gray Hill, who, having tried four times before this to get to Petra from the north was successful in this, his fifth attempt, in April 1896.[23] The atmosphere in Petra had now changed and neither he nor succeeding visitors had trouble there.

During the first few years of the 20th century Sultan Abdul Hamid's dream of linking Damascus by railway to Mecca and Medina was on the way to realisation, and by 1904 the line had reached Maan. Now one could travel by rail to Maan and ride to Petra from there in a few hours. During the 1890s and the first decade of the 20th century European archaeologists and epigraphers were for the first time able to examine Petra thoroughly, with few difficulties.

Such peaceful progress was not to last. This corner of the world, like so many others, would soon be involved in the Great War. The only name in the list of visitors to Petra in the fatal year of 1914 is that of a young archaeologist who would soon be much concerned with Aqaba and the Hejaz Railway – his name, of course, was T.E. Lawrence.

# REFERENCES

1. The information in this chapter is drawn largely from a hundred or so published or unpublished accounts of journeys to Petra in the century between 1812 and 1914. If I attempted to support every statement made above by reference to specific sources, the list of notes might become almost as long as the chapter itself. I have therefore cited books or other sources only when a quotation or author's name appears in the text and I have written references in severely abbreviated form. In addition, however, I have produced a much more detailed annotated chronological list of all the narratives of journeys to Petra between

1812 and 1914 which I have been able to find. This is in typescript; interested readers may obtain a copy from me, N.N. Lewis, c/o ASTENE, 26 Millington Road, Cambridge CB3 9HP.

I have deliberately not used a modern system of transliteration from Arabic to English but have left names in much the same forms as those commonly employed by most of the travellers.

2. J.L. Burckhardt, *Travels in Syria*, London: Murray, 1822, pp. 418-434.
3. C.L. Irby and J. Mangles, *Travels in Egypt*, London, privately printed, 1823, pp. 331-486.
4. L. de Laborde and L. de Bellefonds, *Pétra Retrouvée*, (eds. C. Augé and P. Linant de Bellefonds), Paris: Pygmalion, 1994.
5. J.L. Stephens, *Incidents of Travel*, London: Bentley, 1838.
6. C. and R. Rowley and others. Unpublished material; see the list referred to in note 1 above.
7. Lord Lindsay, *Letters on Egypt*, London: Colburn, 1847.
8. J. G. Kinnear, *Cairo, Petra and Damascus*, London: Murray, 1841; D. Roberts, *Eastern Journal*, manuscript and typewritten copies in the National Library of Scotland, Acc.7723/2.
9. R. Pococke, *A Description of the East*, London, printed by W. Boyer for J. and P. Knapton and others, 1743, p. 138; [J.L. Porter], *Murray's Handbook for Travellers in Syria and Palestine*, 1858, p. 40
10. H. Formby, *A visit to the East*, London: Burns, 1843, p. 254
11. Lindsay, op. cit., 232. Cf. H. Martineau, *Eastern Life*, London: Moxon, 1848, III, pp. 33 and 41. The most serious incident involving visitors to Petra took place half a century later, in 1897, when Lagrange and his companions were ambushed near the south end of the Dead Sea (some 60 or 70 miles from Petra) as they were returning from Petra to Jerusalem. Their attackers were 50 or so bedouin from several different tribes, not including the Alouin, armed with modern rifles, who robbed them and killed two men from Hebron who were travelling in their rear. *Revue Biblique* 1898, Paris: Lecoffre, for the Dominican Convent of St Etienne in Jerusalem, p. 167.
12. E. Robinson, *Biblical Researches*, London: Murray, 1841, II p. 143.
13. The Bdoul numbered about 140 tents or families in 1985; they were almost certainly far fewer in the 19th century.

14. Some recent writersm as well as some of the 19th century, travellers have described Abou Zaitoun as sheikh of the Bdoul. A careful reading of almost all the 19th century accounts has led the present writer to conclude that this was not the case and that Abou Zaitoun was in fact the sheikh of the Liathna, as stated above.
15. E. Lear, 'A leaf from the journals of a landscape painter', ed. 'F.L.', *Macmillan's Magazine*, LXXV, 1896-97, p. 428.
16. M. Plumley: *Days and Nights in the East*, London 1845, p. 121; H. Formby, op. cit., 149-150.
17. D.S. Dodge: letter in *The Evangelist*, New York, 23 April 1868; F.E. Church: letter to E.D. Palmer, 10 March, 1866, in G. L. Carr: *Frederick Edwin Church*, 2 vols., Cambridge U.P., 1994, I, p. 389.
18. J. Loftus: *Mafeesh, or Nothing New*, London, printed for private circulation, 1870, I, p. 199; W. de Famars Testas, 'Journal' in *Album de Voyage; des artistes en expedition au pays de Levant*, Y. Fischer et al., eds., Paris: AFAA/RMN, 1993.
19. de Famars Testas: op. cit., 145-150; J.L. Gérome, Notes de Voyage of, in Ch. Moreau-Vauthier: *Gérome, Peintre et Sculpteur*, Paris, Hachette, 1906, 247-249; P. Lenoir, *The Fayoum or Artists in Egypt*, London: H. S. King & Co., 1873, 272-273, pp. 281-286.
20. J. Finn: *Byeways in Palestine*, London: Nisbet, 1866, p. 289f.
21. Op. cit., 265.
22. A. Hornstein, *Palestine Exploration Fund Quarterly Statement*, London, 1898, pp. 94-100; A. Forder: *Petra: Perea Phoenicea*, London and Edinburgh, Marshall, n.d. 1-62
23. Gray Hill in PEFQS 1897, pp. 33-44, 135-144.

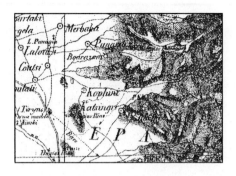

## Chapter Six
# Surveying the Morea: The French Expedition, 1828-1832

## Malcolm Wagstaff

The publications of the *Commission Scientifique de Morée* – an atlas and 10 volumes of text and illustrations[1] – are folio-sized, handsome rivals to those of the better-known, *Commission d'Egypte*.[2] Published between 1831 and 1838, they contain valuable information on the geology and natural history of the Morea (Peloponnese), as well as its then surviving ancient buildings. They are the results of collaboration between the scientists and scholars of the *Commission* and the *ingénieurs-géographes* (geographical engineers) of the army of occupation based in the region at the same time.[3] This contribution will focus on the military surveyors. They produced a remarkable map of the Morea and an enumeration of its population. I will first sketch in the background to their work and then summarise their activities.

*Background*

In reviewing the background to the French Expedition to the Morea, it is useful to adopt both a long perspective and a proximate view. As with Egypt, French governments had had their eye on the region for much of the 18th century.[4] Not only would its acquisition put pressure on the Sultan to support French geo-political objectives in the eastern Mediterranean, it would also consolidate French dominance over the trade of western Europe with the Levant. In addition, the peninsula was seen as having considerable economic potential – arable land, timber and a favourable climate – and might have been viewed as some compensation for the losses of colonies in North America.

Realisation of French ambitions came close under Napoleon. Following the peace of Campo Formio (1798), France annexed Venice and its overseas territories. The Ionian islands, some of which lie off the western and southern coasts of the Morea, were occupied. Although the islands were subsequently lost and given autonomy under Ottoman and Russian protection, France regained them in 1807. The threat of an invasion of the mainland, accompanied by an uprising of the oppressed Christians, was used to break the anti-French alliances formed between the Ottoman Empire, Britain and Russia. The invasion never happened, but the seriousness of the intention is clear. British governments took various counter measures, while in 1807 (perhaps the most dangerous time) the French army urgently commissioned a map of the Morea from the leading French expert on the geography of the region, Jean-Denis Barbié du Bocage (1760-1825), an employee in the Archives Division of the Ministry of Foreign Affairs.[5]

From a more proximate view, the War of Greek Independence provided the opportunity to send an expedition to the Morea. The Christians in much of what is now Greece rose against the Ottoman authorities in 1821.

Despite the insurgents' early successes, a military and diplomatic stalemate developed. To break it, the Sultan enlisted the help of his viceroy in Egypt, Muhammad Ali, an Albanian from Kavalla. Muhammad Ali had the advantages of a European-trained and officered army and of a modern fleet. Some 30,000 Egyptian troops landed in the Morea in February 1825, under the command of Muhammed Ali's son, Ibrahim, fresh from his success against the Wahhabis in the Hejaz and Najd districts of Arabia.[6] Until this point, the other great powers of Europe had regarded the Christian uprising in Greece as a matter for the Sultan, but reports of the atrocities committed by Ibrahim's troops changed attitudes and led to diplomatic and naval intervention. Naval activity to prevent the deployment and re-enforcement of the Egyptian army created that 'untoward event', the battle of Navarino on 20 October, 1827. At least 14 of the 20 ships in the Turko-Egyptian line of battle were sunk or subsequently destroyed.[7] The result was that the Egyptian army was stranded in its camp at Yialova, on the Bay of Navarino.

The idea of sending allied troops to the Morea seems to have been raised first by Lord Palmerston in November, 1827.[8] The French embraced the idea enthusiastically in January-February, 1828 and began to form the initial army corps in March. A protocol signed by the representatives of Britain, France and Russia on 19 July formally invited the French king, Charles X, to provide the army.[9] A force of 14,000 officers and men (infantry, cavalry, artillery including a siege train, engineers and gendarmerie) reached the Morea in late August and early September 1828, and was based at Modon. Its commander was Lieutenant-General Le Marquis N-J. Maison (1771-1840), a distinguished soldier who had served France during the Revolutionary and Napoleonic Wars. Ibrahim Pasha's army withdrew on 16 September and 5 October.

As the military plans matured during the summer of

1828, the idea emerged in court circles of sending a scientific mission to study the Morea.[10] While the inspiration was the work of the Commission d'Egypte, still in the course of publication, the region was attractive for research on three counts. First, it would bring lustre to the restored Bourbon regime, still living in the shadow of Napoleon. Secondly, research would go some way towards realising French aspirations in the eastern Mediterranean, and there was the possibility at the time that French forces in the Morea might stay indefinitely. In that event, first-hand information would prove extremely useful. British pressure prevented a lasting occupation.[11] Most of the French army withdrew in June, 1829, leaving behind only a single brigade; that departed in August, 1833. Thirdly, the Morea was attractive as part of that ancient Greek world which exercised such a fascination over early nineteenth century Europe. A scientific mission could expand the information available to contemporary scholars, artists and architects from the more sporadic observations of earlier travellers.

The Ministries of War and the Navy approved the idea. The Institut de France drew up the instructions and recommended personnel, mostly in their twenties and thirties. Overall responsibility was given to the former Bonapartist, celebrated naturalist, traveller and cartographer, Colonel J.B.G. Bory de Saint-Vincent (1778-1846).[12] His staff was grouped into three sections responsible for architecture and sculpture, archaeology, and the natural sciences. In the event the archaeology section never went into the field, while the work of the natural sciences team was extended to cover places and people.[13] It was at this point that the decision seems to have been made to produce a detailed map of the Morea.[14]

Maps of the region already existed.[15] Barbié du Bocage's map was published in 1814 but, like its predecessors, it was incomplete and inaccurate. A better one was already being compiled by Colonel Leake (1777-1860), a recently retired British artillery officer who had travelled in the Morea during 1805 and 1806.[16] Like the French, Leake knew that a really accurate map of the region would have important benefits, both practical and scholarly. Ancient sites, known from literary sources and archaeological remains, would be fixed more precisely than had been possible hitherto. Modern settlements would be located, providing a secure basis for the study of the population. Accurate information on topography and routes would assist administrators, soldiers and merchants.

A French military surveyor, Captain E. Peytier, had been at work in the Morea since May 1828. He had been sent out in response to a request from the President of the Provisional Greek Government, Ioannis Capo d'Istria (1776-1831), an enlightened Ionian nobleman with a reputation as an administrator and reformer who had recently been joint Foreign Minister to the Tzar and was now intent on creating a modern Greek state.[17] The French War Ministry decided that Peytier should be joined by Lt. Emile Le Puillon du Boblaye (1792-1843) and a team of *ingénieurs-géographes*. They were expected to work in close association with the Commission Scientifique.[18]

*The Survey*

The surveyors and the members of the Commission reached Navarino on 3 March, 1829 and set up their base at army headquarters in Modon. The town was in ruins. There was no inn and only about 100 houses were habitable. However, two or three hundred tradesmen had opened shops and cafes to supply the needs of the French.[19] Colonel Trézel, Deputy Chief of Staff to the army in the Morea, was put in charge of the surveyors. They started work on 1 April 1829 and finished by 10 May 1832. The Commission Scientifique also completed its

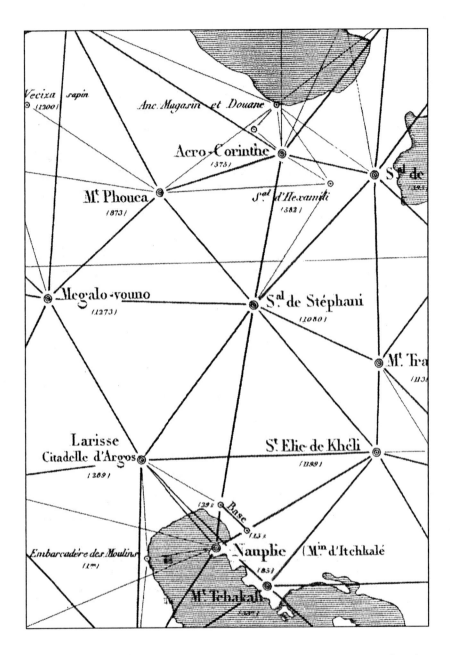

**Figure 22:** *An extract from the* Carte de la Morée *by Jean-Denis Barbié du Bocage, drawn and engraved in 1807 but not published until 1814.*

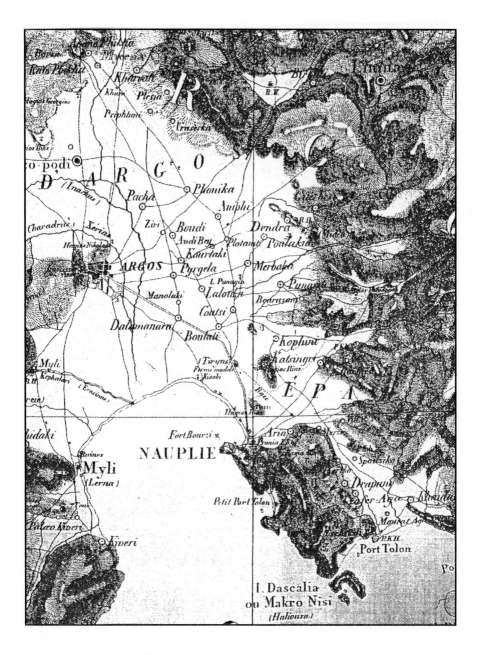

**Figure 23:** *An extract from the* Carte de la Morée *produced by the Geographical Engineers attached to the French Expedition to the Morea and published in 1832.*

work very quickly and landed back at Marseille on 1 January 1830.[20]

The survey of the Morea had two aspects.[21] The most fundamental was the triangulation of the peninsula. To create an accurate map it is necessary to establish a structure of fixed points, to which the detail can be fitted. In the Morea the positions of the necessary fixed points were determined partly by astronomical observations to establish longitude and latitude, but largely by triangulation. Triangulation is a system by which an area is covered with a network of triangles.[22] It depends upon the principle that, when the length of one side and the size of two angles are known, the length of the other two sides and the size of the third angle can be calculated using trigonometry. It is thus possible to fix the location of the points which lie at the angles of each triangle. However, it is necessary to measure the length of one side of an initial triangle as precisely as possible to create an accurate base for constructing the whole network. The triangulation technique was already well-established in the survey of France by the time the Expedition went to the Morea,[23] and it was well adapted to the mountainous terrain and communications difficulties of the region. While the measuring of distances accurately with chains and tape measures was always problematic, angles were easy to measure using a theodolite.

The base for the triangulation of the Morea was measured by Peytier in the Plain of Argos. The great triangles were then created with assistance from Puillon de Boblaye and Servier. In all, some 350 first and second order triangles were created, and about 400 geographical points accurately fixed (Fig. 2). The complete results were published by 1835.[24]

The second aspect of the survey of the Morea was to add topographical detail to the basic triangulation structure. This task was carried out by five teams, each assigned to a different area of the Morea. They consisted of two or three geographical engineers, an interpreter and a small number of sappers. Standard military techniques were employed, together with the mapping conventions adopted in 1803 for all French military survey work.[25] The plane table was the basic instrument. This makes use of a ruler with a sight-vane at each end and a drawing board mounted horizontally on a tripod. The plane table was set up at various points, its position established by resection, ie. observations to positions already fixed by triangulation. Base lines were drawn to scale on paper attached to the board and then the topographical detail was sketched in, using intersecting lines drawn with the sight ruler, as appropriate. The whole procedure was very quick and remarkably accurate.[26] However, the work of survey was beset with problems.

To begin with, the scale of the task was considerable. The Morea has an area of 21,330 km, much of it mountainous, and the number of surveyors was small. Perhaps no more than 10 officers were in the field at any one time. An early decision to reduce the scale of mapping simplified the task. Secondly, all the surveyors were ill at some stage. Of the 18 officers successively engaged in the work, three died and 10 were invalided home. Activity had to be suspended altogether in the summers of 1829 and 1830. Thirdly, competing tasks were assigned to the teams. Bory de Saint-Vincent, for example, had his own ideas over priorities and the nature of the collaboration with his Commission Scientifique[27] while the Greek Provisional Government required detailed mapping of some areas and towns, as well as assistance in checking the population data which it had been gathering.[28] In 1830, the French government itself removed Peytier and the head of field survey, Barthélemy, to serve on the joint commission delimiting the northern frontier of the new Greek state.

Finally, the surveyors were working in a difficult social environment. Atrocities had been committed on both sides during the war of independence. The Muslim population had been eliminated, while many of the surviving Christians had sought refuge away from their homes and were starving. Towns and villages had been destroyed. Although attempts at relief had been mounted from outside, foreigners were not necessarily welcome, soldiers perhaps particularly. In any case, the ordinary people were suspicious of both surveying activity and questions about their numbers. There was also friction between different political factions in newly liberated Greece. Capo d'Istria was assassinated (on 6 October, 1831) and civil war broke out. Barthélemy at one time was put in charge of a force sent to quell rebels at Kalamata.

*Conclusion*

Despite the difficulties, the survey of the Morea was completed by 10 May 1832. Drawing-up must already have been underway, for the *Carte de la Morée* had been engraved and published at the scale of 1:200,000 by the end of the year (Fig. 23). Criticism came from Bory de Saint-Vincent who, as head of the historical bureau in the War Ministry since his return from the Morea, wanted to be put in charge of producing it.[29] Colonel Leake was more generous. His own map of the Morea appeared in 1830. It was based on 'more than fifteen hundred measurements with the sextant and theodolite, made from every important geographical station',[30] and his timed itineraries, as well as information from other recent travellers and British Admiralty charts. When Leake came to publish his 1846 supplement to his *Travels in the Morea*, he adopted the French map, reducing it for convenience of publication and following a different system for rendering place names, but modifying the detail only to a small extent.[31] His redrawing seems to have been completed by 1840 ('drawn and engraved by W. Hughes, 1840'). The French map is a typical military product of its time, showing topography with hachures, marking the roads and locating settlements which are represented in an hierarchy of *villes*, *bourgs* and *villages* (Fig. 23).[32]

As to the population data collected jointly by the Commission Scientifique and the surveyors, a number of deficiencies emerge when they are examined. The notes to the tables published by Bory de Saint-Vincent show that information was not always forthcoming from local leaders and older estimates were employed instead.[33] In half of the 28 administrative districts most settlements returned the number of individuals, as well as the number of families, but in the other half only the number of families was provided. Many villages, however, were named in the tables, but without any population being noted. In other cases, the populations of neighbouring villages were aggregated. Comparison of the population tables with the maps shows further that some villages were omitted altogether from the enumeration.

On the other hand, the population data are reasonably comprehensive for the whole region and, at the level of the number of families, broadly comparable with those from the Grimani Census of 1700, taken during the Venetian occupation of the Morea.[34] Moreover, since the data are given for named settlements, they can be linked to the French map, allowing notions of pattern, density and distribution to be formed. In other words, from the French map and the enumeration it is possible to create a geography of the Morea as it was perceived by the French Expedition. The moment was critical in the history of modern Greece: Ottoman rule had ended and Greek independence had been recognised.

# REFERENCES

1. Expédition Scientifique de Morée, Paris et Strasbourg: F.G. Levrault:
   T. 1, Relation du Voyage, 1836
   T. 2, Pt. 1, Géographie , 1834
      Pt. 2, Géolgie et Minéralogie, 1833
   T. 3, Pt. 1, Zoologie, 1832
      Pt. 2, Botanique, 1832
   *Monuments d'Antiquité, recueillis en Grèce par la Commission de Morée*, Paris: F.G. Levrault, 1835.
   Expédition Scientifique de Morée, *Recherches géographique sur les ruins de la Morée*, Paris: F.G. Levrault, 1836.
   Commission Scientifique de Morée, *Section d'Architecture et de Sculpture*, 3 vols., Paris: F.G. Levrault, 1831-39.
2. Jomand, E.F. et al., *Description de l'Egypte*, 9 vols. text, 11 vols., plates and three over-sized vols, including an atlas, Paris: Imprimerie Imperiale/Royale, 1809 (1810)-1828.
3. Crawley. C.W., *The Question of Greek Independence: A Study of British Policy in the Near East, 1821-33*, Cambridge: Cambridge University Press, 1930, p. 398.
4. Naff, T., *Ottoman Diplomacy and the Great European Powers, 1797-1802*, unpublished PhD Thesis, University of London 1960, p. 256, n.46
5. Barbié du Bocage, J-D., *Carte de la Morée dresée et gravée au Dépôt Géneral de la Guerre par Ordre du Gouverment en 1807*, Paris: Imprimerie Imperiale/Royale, 1814. See also Whitcomb, E.A., *Napoleon's Diplomatic Service*, Durham N.C.: Duke University Press, 1974, p. 174.
6. Crabitès, P., *Ibrahim of Egypt*, London: George Routledge and Sons, 1935; Driault, E., *L'Expédition de Crète et de Morée (1823-28): Correspondence des Consuls de France en Egypte et en Crète*, Cairo: La Société Royale de Géographie d'Egypte, 1930.
7. Anderson, R.C., *Naval Wars in the Levant, 1559-1853*, Liverpool: Liverpool University Press, 1952, p. 532.
8. Crawley, op. cit.,102, n.5, 107-108.
9. The account of the formation of the French expeditionary force and its early progress is based upon Baloti, Z., 'La formation du Corps

*expédionnaire française de Morée en 1828'*, Deltion tis Istorikis kai Ethnologikis Etaipeias tis Ellados 32 (1989) pp. 93-108.

10. Broc, N., 'Les grandes missions scientifiques françaises au xixe siècle (Morée, Algérie, Mexique) et leurs travaux géographiques', Revue d'Histoire des Sciences, 34 (1981), 318-58; Bory de Saint-Vincent, J.B.G., Relation du Voyage de la Commission Scientifique de Morée dans le Péloponnèse, les Cyclades et l'Attique, 2 vols., Paris: 1836
11. Broc., *ibid.*; Crawley, op. cit., pp. 157-58.
12. *Nouvelle Biographie Générale*, Copenhagen: Rosenkilde et Bagges, T. 6, 1964, pp. 750-754.
13. Bory de Saint-Vincent, op. cit.
14. Berthant, H. H., *Les Ingénieurs Géographes Militaires, 1626-1831*, T. 2, Paris: Service Géographique de l'Armée, 1902, pp. 464-76.
15. For example, d'Anville, J.B.B., *Les Côtes de la Grèce et l'Archipèl*, Paris: Imprimerie Royale, 1756; d'Anville, J.B.B., *Alalyse de la Carte Initulée les Côtes de la Grèce et l'Archipèl*, Paris: Imprimerie Royale, 1757.
16. Leake, W. Martin, *Travels in the Morea*, London: John Murray, 1830, Vol. 1; Leake, W. Martin, *Peloponnesiaca: A Supplement to Travels in the Morea*, London: J. Rodwell, 1846, pp. v-ix.
17. Woodhouse, C.M., *Capodistria: The Founder of Greek Independence*, London: Oxford University Press, 1973, pp. 387-431.
18. Berthant, op. cit., pp. 464-66.
19. Bory de Saint Vincent, op. cit., pp. 60 and 97.
20. Bory de Saint-Vincent, *op. cit.*
21. This account of the work of the surveyors is based upon Berthant, op. cit., pp. 464-76, *unless otherwise stated.*
22. War Office, *Manual of Map Reading and Field Sketching*, London: His Majesty's Stationery Office, 1921, pp. 74-78, 179.
23. Godlewska, A., 'Map, text and image. The mentality of enlightened conquerors: A new look at the Description de l'Egypte', Transactions of the Institute of British Geographers, NS., 20 (1995), 5-28 (n. 70).
24. Commission Scientifique de Morée, *Triangulation et Levées de la Grèce par l'Etat-Major Française*, Paris: 1830-35
25. Godlewska, op. cit.
26. War Office, op. cit., pp. 94-120.
27. Bory de Saint-Vincent, op. cit.

28. For the census, see 'Expedition Scientifique de Morée', T. 2, Pt. 1, *Géographie, pp. 60-94. On the mapping of towns see 'Oi martyries perihghton kai apestalmenon gia tis poleis tis Othomanikis Peloponnisou kai topographikes apotyposeis tis peridiou 1828-1836' [Evidence from travellers and officials for the towns of the Ottoman Peloponnese and the topographical prints of the period 1828-1836], In, Kalligas, H.A. (ed.), Travellers and Officials in the Peloponnese, Monemvasia (Greece): Monemvasiotikos Homilos, 1994, pp. 186-222.*

29 Expedition Scientifique de Morée, op. cit., T. 1, 1836, p. 244; T. 2, Pt. 1, *1834, p. 52.*

30. Leake, W. Martin, *Travels in the Morea*, London: John Murray, 1830, *Vol. I, p. vii.*31, Leake, W. Martin, *Peloponnesiaca*, London: J. Rodwell, 1846.

32. Comparison is invited with the early maps produced by the Ordnance *Survey in Britain. See Harley, J.B., The Historian's Guide to Ordnance Survey Maps, London: Standing Conference for Local History, 1964; Seymour, W.A. (ed.), A History of the Ordnance Survey, Folkestone: Dawson, 1980.*

33. Expedition Scientifique de Morée, op. cit., T. 2, Pt. 1, pp. 60-94.

34. *Libro Ristretti delle fam[i[g[li]e et anime efftive in cadaun territori[o] del Regno, Venice, Archivio di Stato, Archivio Grimani dai Servie, Busta 54, No. 158.*

# SELECT BIBLIOGRAPHY

Baloti, Z., 'La formation du Corps expeditionnaire française de Morée en 1828', *Deltion tis Istorikis kai Ethologikis Etaipeias tis Ellados*, 32 (1989), pp. 93-108.

Berthant, H.H., *Les Ingénieurs Géographes Militaires, 1626-1831*, Paris: Service Géographique de l'Armée, 1902.

Broc, N. ' Les grandes missions scientifiques françaises au xix siècle (Morée, *Algérie, Mexique) et leurs travaux géographiques'*, *Revue d'Histoire des Sciences*, 34 (1981), pp. 318-358.

Crabitès, P., *Ibrahim of Egypt*, London: George Routledge and Sons, 1935.

Crawley, C.W., *The Question of Greek Independence: A Study of British Policy in the Near East, 1821-1833*, Cambridge: Cambridge University Press, 1930.

Driault, E., *L'Expédition de Crète et de Morée (1823-1828). Correspondence des Consuls de France en Egypte et en Crète*, Cairo: La Société Royale de Géographie d'Egypte, 1930.

Godlewska, A., 'Napoleon's geographers (1797-1815): imperialists and *soldiers of modernity'*, In Godlewska, A. and Smith, N. (eds.), *Geography and Empire*, Oxford: Blackwell, 1994, pp. 31-53.

Godlewska, A. 'Map, text and image. The mentality of enlightened *conquerors: A new look at the Description d'Egypte'*, *Transactions of the Institute of British Geographers*, NS., 20 (1995), pp. 5-28.

Jones, Y., 'Aspects of relief portrayal in nineteenth-century British military maps', *Cartographic Journal*, 11 (1974), pp. 19-33.

## Chapter Seven
# *La Mission Scientifique de Morée: Captain Peytier's Contribution*[1]

# Elizabeth French and William M. Frick

The French Mission Scientifique contributed significantly to the understanding of the Morea. One of the players in the drama of the Morea investigations was Captain Jean Pierre Peytier.

The fact that he was already in Greece when he was transferred to the Expedition Scientifique in January 1829, as well as the quality of his work and the interest of his observations, make him a particularly intriguing subject of study.

His career[2] illustrates the principles of the French Revolution into which he was born. Jean Pierre Eugene Felicien Peytier was born in the village of Sandrons in the Ardèche, probably on 15 October 1793, to a young and uneducated farmer who could do no more than add a signature to his son's birth certificate – it had to be attested by a local magistrate. From Peytier's service record we learn that the boy was an *élève* in the École Polytechnique in September 1811, ranking fourth out of 175. In September

1813 he was first of 21 as a 2nd lieutenant in the Geographic Engineers. He served briefly at Tours during Napoleon's 100 days and then returned to the Dépôt de la Guerre to begin work as a cartographer, mapping Paris. After further study he began, in April 1818, his life's work of mapping France. As far as the authors know, no picture of him exists.

Ten years later, in February 1828, he was sent with another engineer and two artillery officers, at the request of Capodistrias, to organise a Greek Officer Cadet School[3] and to undertake other duties.[4] Capodistrias had himself returned to Greece only in January 1828, once the battle of Navarino in October had opened the way, though he had been appointed *kyvernitis* by the Assembly at Troizene in April 1827. His background at the Tsar's court, and the distorted perspective of distance, might have made it seem a good idea to train military cadets in a western system, but it proved an almost impossible task. From the military point of view, few of the Greek irregulars were yet ready for such discipline.[5] Moreover the political situation was highly unstable, as Capodistrias was quite unwilling to conform to the limitations imposed on him by the Troizene assembly. He dismissed that assembly, appointing in its stead first a group of 27 members under his personal control (the Panhellenion), and then, in July 1829, a fourth national assembly which convened in the theatre at Argos.

While Capodistrias at first welcomed Peytier and his colleagues, in time his government's failings and doubts about French intentions cooled the relationship, to the extent that all four had only unimportant assignments. The French Military Mission, which was to clear the Turks from the Morea and establish some form of government, had sailed from Toulon in August 1828 but Peytier and his colleagues were not attached to this. Capodistrias' suspicions may well have been justified: the complement of cartographic engineers and artillery officers seems unnecessarily specific for the tasks in hand. There was apparently nothing unconscious about the selection: the French government was well aware of the disposition of its specialists when it came to finding cartographers to assist the Mission Scientifique slightly later, though their role was carefully defined on both sides.[6]

Early in 1829 the lack of constructive work, together with his own observations, aroused in Peytier the idea that he should of his own volition map the Morea. Apparently he had accomplished some of the work by the time he joined the Expedition. He had spent part of the year 1828-29 based in Corinth and from there sent a seven page handwritten report to the Société de Géographie.[7] His handwriting is not always clear, and neither is his vocabulary, but the report contains much of importance to the general background of the period.

From this account of his work and of the everyday living conditions in the area, we gain a dramatic picture of the state of the Morea during this period in the wake of Ibrahim Pasha's scorched earth policy. This is usefully complemented by the accounts of the young American doctor Samuel Gridley Howe.[8]

The fighting and the absence of any maintenance between 1821 and 1828 had destroyed the infrastructure of the Morea: the rural economy was totally undermined. General Maison, who commanded the Military Mission, rightly told his troops (harking back to the French experience in the dire Fourth Crusade): 'Hardships, fatigue await you, sustain them with courage.'[9]

They faced difficulties at least as great as those of modern 'peacekeepers' and aid workers in the Balkans and elsewhere. The conditions they encountered devastated the health of the Mission Scientifique.

Though the actual Mission Scientifique had left Greece by November 1829, the cartographers remained. Peytier returned to France in 1831, mainly for family reasons, but also to deal with publishing the map. He left his two colleagues, Sevier and Boblaye, to complete the details. Peytier was highly thought of by his superiors, who put their assessments of his work and worth on record.[10] We can contrast this very professional production (see M. Sève, *Les Voyageurs Français à Argos*, Fig. 9) both with the previous sketch maps by Fourmont (1729), Anville (mid-18th century) and Barbié du Bocage (1807)[11] and with the local plan of Mycenae made by the architects Blouet and Ravoisié.[12] The villages and toponyms on the large map are of considerable archaeological interest.

Peytier returned to Greece again in 1833 to map north of the Isthmus. It is to this period that his three articles in the *Bulletin* for 1837 refer. These were on climate, disease and the general background to the topographic work.

Peytier's contribution to the work of the French in Greece, outlined above, and the other information which can be found in his work, have been neglected. This additional information is derived first from his handwritten report of 1829, and secondly from the archive of his drawings.

In Peytier's report he said that he could 'distinguish clearly' the columns of the Parthenon through his telescope and he gave accurate heights for the main mountains of the Peloponnese. There was much detail too about the people, their houses and their living conditions. The kind of simple house he described can be compared with the account and illustration by William Simpson of the *Illustrated London News* some 50 years later.[13] Little seems to have changed in that time.[14] Simpson's engraving shows that the foustanella was still being worn. This form of dress had become a symbol of the insurrection (Howe was painted wearing it) for it had been the ordinary dress of the *klephts* and *pallikaria* whom Peytier was trying to train. It is now, of course, the standard wear of the *evzones* on official duty in Athens at the Tomb of the Unknown Warrior and at the former royal residence and guest house. Peytier alone tells us – no doubt from his own experience with the *pallikaria* – that these garments were 'dry cleaned' every six months or so by being whirled and beaten over an open fire to dislodge the infesting wildlife which themselves became nicknamed *pallikaria*.[15] The only oddity is that he says that the *fustanelle* is an 'espèce de jaquette' rather than the pleated woollen skirt.

In addition to Peytier's reports and articles, some 11 pencil sketches and 69 sepia prints and watercolours by him are preserved in a private collection in Lucerne,[16] representing both periods of his work in Greece. These formed the subject of a handsome volume published by the National Bank of Greece on the occasion of the celebrations of 250 years of Greek Independence.

The range of these illustrations is telling. There are several (two of Patras, and one each of Amfissa and Corfu) which reflect the journeys to and from Greece. Most of the watercolours concern Athens and can be dated on internal evidence to both before and after 1834. These display great topographic accuracy and are important documents of this period of the city's history. Many are studies of people and their varied costumes, in which he was clearly very interested. These include both original drawings and copies from other sources. The style thus varies, but again they are a valuable source of evidence and can usefully be compared with those of Lear on similar themes some 20 years later.

Interestingly, a watercolour of Troizene is also a copy. There is an acknowledgment on it which says that it was taken from a sketch by 'm. de Semanville(?), ensiegne à bord de la Junon(?)'. Undoubted but unacknowledged copies include a series of small views and costume pieces from Constantinople, Adrianople, Smyrna and Cairo, and a series of the monuments of Phyllae. These may have been done in order that he might issue an album of 'oriental travel', as had become fashionable in the salons of Paris.[17]

Figure 24: *'Drilling the Typikon'*.

**Figure 25:** *'Palamide, as from our lodgings (at Nauplion)'.*

As has been noted, some of the sketches of the Greek military are almost caricatures and can probably be dated to the early part of his career when he was in close contact with such people. 'Drilling the Typikon'[18] (Fig. 22) is typical, and there are others of well known persons such as Kanaris and Kolokotronis.

The four watercolours of Nauplion are particularly delightful. They are also intriguing. It is difficult to decide which period of his work they belong to. If they are from his very early period there, before he went to Corinth, they would have been made relatively soon after the exchanges of cannon fire between the two fortresses (the Palamede and the Uçkale) in the summer of 1827,[19] but they appear to show little damage, particularly to the Bouleutikon. Rebuilding may have been undertaken quickly in the potential new capital: certainly many of the buildings now being restored date from this period and from the early years of King Otto. Perhaps the new buildings were inserted into vacant plots. Urban regeneration may well have preceded that of the countryside.[20]

On the other hand the surveyors must have been based in Nauplion again after they joined the Mission. The base line for their whole plan runs along or just east of part of what is now the main road from Argos, near Tiryns. Agape Karakatsane, in her commentary on the paintings, suggests that the Mission hired the former *medrese* building beside the Bouleutikon (now know as the Leonardo), though she gives no source for this information. She suggests that the watercolour titled 'Palamide as from our lodgings (at Nauplion)' was sketched from the top floor window at the north end of the Leonardo.[21] If so, some liberties have been taken in the final presentation. Peytier's picture shows the outer porch and entrance stair of the Bouleutikon intact and crowned with a further series of cupolas. These have since been destroyed[22] but show also in the engraving from the other side of the square by Lange of slightly later date.[23] The low campanile in the foreground at the left is more puzzling, and as yet defies identification. It may have been merely artistic license to give some contrast to the surrounding cupolas. Otherwise the view is accurate, showing both the dome of Hagia Sofia and the tall houses flanking the slope of the Uçkale right up to the Palamede with its flag of free Greece clearly displayed over the entrance.

Another charming picture by Peytier gives a further illustration of the local soldiery leaving the land gate. This monument, which was destroyed some years ago, has recently been excavated and reconstructed.[24] His view of the town from the Pronoia area shows the size of the town at this period: that of the Hourmoudia district is both explicit and probably very accurate, showing as it does the palm tree which, it is claimed, was planted by Capodistrias himself[25] – another of his notable innovations.

Thus, thanks to the keen interest and intelligent eye of this redoubtable geographical engineer, both the report and the illustrations give a valuable understanding of the background against which the scientific work of the Mission was done.

## REFERENCES

1. This paper resulted from a query from Elizabeth French to William M. Frick. In his privately circulated *Life of Col. Steffen*, William Frick quoted from the articles by Peytier in the *Bulletin de la Société de Geographie* about working conditions in Greece, and later he used several of Peytier's watercolours in another monograph. He thanks Col. Serge Gabriel, Mrs George Gekas, Mrs Elberito Scocimara, Mrs Teryn Weintz, and Mon. Olivier Rouer for their help in his researches.

What is presented here represents the current state of our knowledge, but there is definitely more to be discovered. The authors have seen only a small part of the original material.
2. See source list quoted by Papadopoulos, p. 102.
3. This was in the building which has now been refurbished as the War Museum in Nauplion. The current officer cadet school at the east base of Hymettos, still named the Evelpidhon, is the direct legacy of this foundation.
4. Papadopoulos, n.134-5.
5. They appear in several of Peytier's drawings almost as caricatures.
6. Papadopoulos, n. 120 *et seq*.
7. Archives of the Société de Geographie (Sg. Colis. No 19 bis[3225]). More work needs to be done on this document, which is not quoted as a source by Papadopoulos.
8. Howe's papers concerning this period are in the Houghton Manuscript Library of Harvard University. See particularly Files b44M-314 (1403-4), 1421 (typescript) and 1559. His information is clearly used by Jane Aitken Hodge in her historical novel *Greek Wedding* in which Howe is a character and vivid scenes are set in Nauplion in the period just before the battle of Navarino.
9. The letter from Maison dated 13 août 1828, ends: 'Le prevation, le fatigue vous attendent, vous les supporterez avec courage'.
10. Papadopoulos, n.142.
11. All discussed by Sève pp. 38-40.
12. Expédition II, pl. 62.
13. ILN April 21, 1877 p. 364 'Interior at Mycenae'.
14. Simpson at least had a bed in Corinth but he says it was supplied by his guide.
15. *Pallikari* remained the name for a 'young blood' until at least the 1950s though it has given way to a new term now.
16. Elizabeth French thanks Mr Finopoulos of Athens for this information. He kindly assisted when she was unable to obtain any information from the National Bank of Greece. She is currently attempting to contact the remaining members of the family of Stephanos Ballianos who originally bought the archive.

17. See the paper by Anna Piussi presented at the 1999 ASTENE Cambridge Conference.
18. The Typikon was the official army of the new Greek government.
19. The Palamede was held by T. Grivas and the Uçkale by N. Photomara. On 1 July over 200 balls were fired from the Palamede causing considerable damage in the town and panic among the inhabitants. The next day a ball hit the Bouleutikon, landing in the chamber of the Bouleuterion. Gerothanases was killed and another deputy sitting beside him wounded as is attested in a letter written by another deputy who was present. Two days later the rogue American Philhellene William Washington Townsend was killed in the street. Elizabeth French is grateful to Professor I.F. Clarke for taking the trouble to correct her on the subject of 'bombardment' after the original presentation of this paper.
20. Similarly Blouet's commentary on contemporary Argos is highly favourable mentioning that 'each house has its garden; the air of cleanliness that we have found there is not always found in other towns of the Morea. This one has a very beautiful situation.' The engraving of the general setting of Argos in the Mission publication is certainly romantic and contrasts with the serious detail of their architectural drawings.
21. See Fig. 23. The Leonardo has, since the 1960s, been used as storerooms and a study centre for archaeologists working in the Argolid. Elizabeth French has a considerable collection of photographs taken from the roof just above this window (which is now a part of the conservation section of the Nauplion Museum).
22. No source that the authors have managed to consult gives a date for the demolition of these features.
23. Karouzou fig. 69 in the Graphische Sammlung in Munich. Lange came to Nauplion in 1834 with Rottmann.
24. Though without the elaborate forked entrance which shows in the engraving by Karl von Heideck (Karouzou, fig. 106).
25. According to a sign on the trunk.

# SELECT BIBLIOGRAPHY

Karouzou, S., *To Nauplio*, Athens, 1979.

Lambrinidos, M.G., *E Nauplia apo ton Archiaotaton Chronon mechri ton kath' emas:* Istorke Melete, Athens, 1950.

Papadopoulos, S.A. and Basilake-Karadatsane, A., *E Eleuthere Ellas kai e Epistemonike Apostole tou Moreos*: To Leukoma Peytier tes Sylloges Stephanou Ballianou. National Bank of Greece, Athens, 1971.

Sève, M., *Les Voyageurs Français à Argos*, EFA Sites et Monuments X, 1993.

## CHAPTER EIGHT
# Christian Rassam (1808-72): Translator, Interpreter, Diplomat and Liar

# Geoffrey Roper

The family name Rassam is probably best known from the career of Hormuzd Rassam (1826-1910), who acquired some fame in the 19th century, first because of his dramatic adventures as a British emissary in Ethiopia, and then as an archaeologist and Assyriologist. Less celebrated is his elder brother Christian Anthony Rassam, but Christian's life was also full of journeyings and escapades, both physical and intellectual, and he interacted with a number of more famous travellers in the Near East.

The pencil and chalk portrait by William Brockedon, now in the National Portrait Gallery in London, was done in February 1838, when Rassam was in England. It is signed just C. Rassam, the 'C' being short for 'Christian', the anglicised form of his name, which he adopted for most purposes after the age of about 30. But he first becomes known to us under the Arabic form 'Īsá Rassām, which is how he signed the portrait in the Arabic script. Moreover he bracketed Arabic with Kurdish, widely used in his native region, in which his name is spelled in the same way. He also signed himself Yeshū'a Rassām in Syriac, or 'Chaldean' as he called it, the preferred language of his people. 'Īsá and Yeshū'a are forms of the name Jesus, and the family name Rassām means 'draftsman' – more specifically one who designs and traces patterns for textiles. This was his family's traditional trade.

Rassam was a 'Chaldaean' Christian from Mosul. This means that he belonged to that section of the local Nestorian Church of the East which had entered into communion with Rome. That much we know, but as soon as we try to establish anything further about his background and his early life, we are immediately confronted with a problem: all the accounts come from himself, and most of them are mendacious. We know this simply because they differ so much from each other.

To his Anglo-Catholic friend and patron William Palmer he claimed that his great-grandfather was a Spaniard, and that his family were staunchly pro-Papal, occupying positions of influence in the Uniate community.[1] But according to what he told his employer Francis Chesney, his great-grandfather had come from Malabar.[2] According to his brother Hormuzd Rassam, his father was an archdeacon.[3] He himself told his missionary supervisor Christoph Schlienz that his father was a British subject from India, who had settled in Mosul.[4]

He was, it seems, born in 1808: he was at least consistent about that. But there are a number of accounts of his early life and first contacts with Protestant missionaries, differing considerably in their details, but all of them by, or emanating from, the man himself. It seems that he told different stories on different occasions, to suit different audiences. The fullest version is that given in conversations at Oxford with William Palmer, and transcribed by the latter in a notebook now preserved in Lambeth Palace.[5] According to this, he received an elementary education from what he called a 'stupid old priest' in Mosul, and was in turn employed to teach other children. When he was 16, in 1825, he was sent off to be trained as a priest in Rome, and on the first stage of his journey he accompanied a caravan across the Syrian desert to Damascus.[6] But according to the story which he later told his patron, Col. Chesney, he went with the caravan to trade 'in the heart of Arabia';[7] whereas according to the account of himself which he had given earlier to the Secretaries of the Church Missionary Society, he was accompanied not by a caravan but only by 'three monks', and was attacked in the desert by bedouin and stripped of all his possessions[8] – an archetypal tale of misfortune. He had previously told Schlienz that he was accompanied by two monks, and that the attack took place near Aleppo.[9]

From Damascus he went on to Egypt, still intending to proceed to Rome, and visited his uncle, who ran a factory for printing calico in Bulaq. The latter discouraged him from continuing his journey, and instead employed him for 'some years' – so he said – in his factory. While there, he said that he 'heard that there were priests at Cairo selling the S[acred] Scriptures in many language'. He 'went directly to them and bought a Syriac testament and an Arabic Bible & some tracts which were printed in Malta'.[10] According to the version transmitted by Schlienz, he was seeking a Bible for his cousin.[11] He observed that the tracts were in 'very bad Arabic' but he nevertheless cultivated the acquaintance of the missionaries. To Palmer, he later attributed this to mere curiosity about their religious views;[12] but to the missionaries, he had claimed that his 'eyes were opened' and that he had come to 'possess, too, the first thing, the truth of the Gospel'.[13]

He eventually left his uncle's employment and in 1829, at the request of the missionaries, joined the Church Missionary Society (CMS) establishment in Cairo, helping with teaching in the school. He himself learned English here, and improved his literary Arabic. A missionary report from that time commended his 'indefatigable zeal for Philology'.[14] According to what he later told the American missionary Perkins, on the other hand, he 'fell into the hands of [the] missionaries' while 'detained by sickness at Alexandria';[15] whereas according to Rassam's rather gullible trading partner of later years, Henry Ross, he was studying at Al-Azhar, no less.[16]

The precise origins and identity of Rassam seem to have been unclear to the CMS missionaries themselves at this stage, and indeed for some time afterwards, because in their early references to him they call him 'Isa Baghdadi' or 'Isa of Baghdad' – it seems, therefore, that he may have told them that this was where he came from. Confusion has also arisen because in later years at least two of his distinguished English acquaintances, Henry Layard and Edward Mitford, referred to him as an Armenian – that is what he probably told them.[17] By then he was generally known as 'Christian Rassam' in European circles.

But however that may have been, he impressed the missionaries with his linguistic and literary abilities sufficiently for them to employ him in the work of translating and revising Arabic tracts.[18] They also formed a good opinion of his moral character, and, according to a letter in the missionary archives dated 1830, 'never have we observed in him the Arabic love of money'.[19] Presumably they were not yet aware of his economy with the truth, or rather, perhaps, his profligacy with the less than true.

In 1830 it was decided to send him to Malta,[20] which was the main base of the Mediterranean Mission of the CMS, and the place where most of the work of preparing and printing missionary books was carried out. By May 1832 he was at work there, and the Superintendent of the Mission, Christoph Schlienz, reported that 'the Arabic youth Isa Rassam of Baghdad has proved hitherto very much to our satisfaction. Although he has not so much knowledge of the Arabic as Fares had, yet he has a clear mind, is attentive to his business & labours more than he is expected'.[21] In the following year, Schlienz was instructed by the CMS Secretary in London to ensure that Isa's services were retained. In his words: 'The openings in the Mahomedan countries in the Levant render it important to place the Arabic Department of your Press on as efficient a footing as possible. For doing this Isa seems well qualified.'[22]

In the latter part of 1833 he was called upon to prepare a primer in Syriac for the American missionary Justin Perkins, who called at Malta en route to Kurdistan. This he did, using his own handwriting, reproduced by lithography, because no Syriac types were available.[23] In March 1834 the famous missionary traveller Joseph Wolff found this 'excellent Chaldean', as he called him, still in Malta 'actively engaged in the missionary arsenal... preparing... mighty engines for conquering the Kingdom of the Devil'.[24] His permanent appointment was confirmed in the same month, after Rassam had given an account of himself and signed a declaration that he would be content with a salary 'only enough for my living', and that his mind was 'entirely bent on this work and I wish no other employment as long as God spares my strength and my life'.[25] So he seems to have settled down in his work at the press, being reported as still making good progress and rendering great help with the Arabic texts in November 1834.[26]

One of the Arabic books published by the CMS in Malta at this time was Bunyan's *Pilgrim's Progress*, which appeared in 1834 under the title *Kitāb Siyāḥat al-Masīḥī*: (see Fig. 26). Rassam later claimed to Palmer that he had done the translation[27], although it has generally been credited to Schlienz: probably they both worked on it. Rassam also claimed to have revised the Arabic version of the Anglican *Book of Common Prayer*: this was published

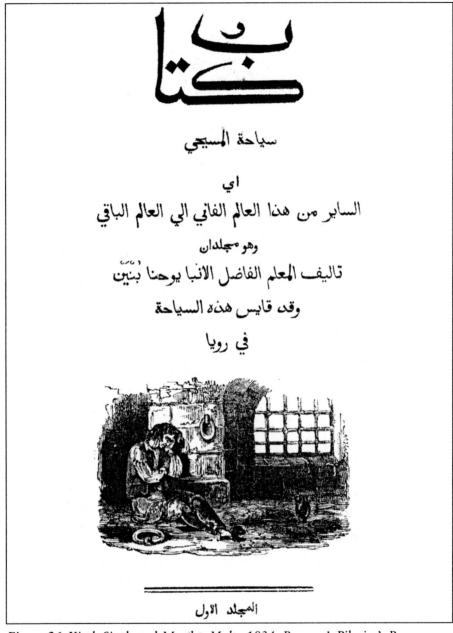

*Figure 26:* Kitāb Siyāḥat al-Masīḥī: *Malta 1834. Bunyan's* Pilgrim's Progress, *translated into Arabic by Christoph Schlienz and Christian Rassam. The wood-cut vignette depicts the author languishing in Bedford jail.*

in Malta in 1840, and Palmer lamented that Rassam had not received any of the credit for it;[28] but according to the CMS archives, this revision was carried out in 1836, after Rassam had left the CMS, by Schlienz and Samuel Lee in Cambridge, together with Faris al-Shidyaq in Malta.[29] Rassam had, however, prepared specimens in 1834, under Schlienz's supervision, which had been sent to the SPCK,[30] and it was presumably to these that he was referring. They seem to have been rejected.

According to his own later account, during this time Rassam became increasingly disillusioned with the CMS and with the ideas contained in the religious texts which he was called upon to translate. Many of these tracts, Palmer reported him as saying,

> contained doctrines quite contrary to those which he learned when a child, especially about Baptism, and about the Millennium. One Catechism which he translated was full of the Millennium, till it made his heart sick, & he thought it would be better to go & live in his own country & be a slave under the Turks than engage himself to assist the Missionaries in spreading poisonous doctrines.[31]

In fact, examination of the Arabic tracts and catechisms on which Rassam worked during this period reveals comparatively little about this subject. It seems that in this matter too he may have been guilty of some disregard for the truth, or at least of very considerable exaggeration, in the presence of his Anglo-Catholic patron.

What was rather more true was another remark which Rassam made later to Palmer, that the missionary work in the Mediterranean had become what he called a 'business'. 'The Missionary,' he said, 'makes a kind of trade of writing and translating the tracts & books: for otherwise the inutility of their mission would be too evident.'[32] But this 'trade' afforded Rassam stable employment with which he declared himself quite content at the time, according to the reports in the missionary archives.[33]

However that may be, when Rassam was presented with an opportunity to go back to 'his own country' and abandon the work on which he had previously pledged that his mind was 'entirely bent', he eagerly took it. This occurred in March 1835, when Colonel Francis Chesney arrived in Malta.[34] He and a team of officers and engineers were on their way to Syria to launch the famous Euphrates Expedition, which was to involve surveying the rivers of northern Syria and Iraq in steamboats, with a view to establishing a new overland route to India, and a means of deterring a possible Russian invasion of the area. Chesney wanted an interpreter, and Rassam, being a native of the area, and having by now a good knowledge of English as well as Arabic, seemed well qualified for the task. He accordingly invited him to join the Expedition, and Rassam, according to Chesney, 'enthusiastically quitted his position'.[35]

The missionaries, however, reported that Rassam had 'yielded only with the greatest reluctance' to Chesney's 'urgent and pressing entreaties'.[36] It seems that he was mainly worried about his continued employment, or re-employment, after the end of the expedition, and that adequate provision should be made for his wife. For at the beginning of that year, 1835, he had married Matilda Badger,[37] the sister of the English missionary, traveller and scholar George Percy Badger. Both these matters were resolved by the intervention of the Governor of Malta, Sir Frederic Ponsonby, who arranged an allowance for Matilda while Rassam was away, promised a pension in case of his death, and pledged to do something for him if he did not eventually resume his missionary employment.[38] How this last promise was redeemed we shall shortly see.

Twelve Maltese were also, in Chesney's words, 'engaged under him to facilitate... communications with the Arabs and be generally useful',[39] and Chesney and Rassam sailed from Malta on 20 March 1835.[40] Rassam served with some distinction on the Euphrates Expedition, acting as 'purser as well as interpreter'.[41] Certainly Chesney formed a high opinion of him, referring to his 'zeal, superior intelligence and usefulness,'[42] and he strongly commended him afterwards to his acquaintances in England.[43] Rassam, for his part, according to another member of the expedition, 'expressed his satisfaction at being set free from the strict rule of the pious folk at Malta'.[44]

The expedition,[45] landed in northern Syria and lugged their baggage overland to the northern Euphrates. Rassam had the responsibility of dealing with all the local helpers, bystanders and suspicious populace at each stage of the journey. The steamers were assembled at a specially built shipyard on the Euphrates, which Chesney called 'Port William', in honour of the King. He soon realised that his original schedule was wildly over-optimistic, and that he would need at least an extra year to complete the project. So he wrote to the CMS, expressing his desire to retain Rassam, whom he described as a 'single hearted zealous and indefatigable man' as his interpreter. He promised to 'restore him after I have gone up the Tigris, if such be the decided wish of the Society', but also stated his 'anxious wish to retain him... for 12 months hence'.[46]

In the event he did retain him, despite Badger's plea to Rassam to return to Malta.[47] No doubt Badger was thinking of his sister, Rassam's wife, who had been left behind, as much as of the Arabic work which remained to be done there. The latter problem was solved by the permanent employment of Fāris al-Shidyāq,[48] a much better translator than Rassam, and in view of this the CMS reluctantly agreed to an extension of Rassam's stay in Iraq.[49]

Eventually the two steamboats set off downstream. One of them disastrously sank in a storm: fortunately for Rassam, he was on the other one. Some of the Arab tribes were friendly, others less so. Rassam had to negotiate with them for food supplies, which he did with great success. The party regularly dined off whole sheep, and it was recounted by one colleague that Rassam always insisted on keeping the liver for himself, which he shared with a female member of the party who was also something of a gastronome. He was also noted for whiling away the time with games of chess, which he always won.

When the main party set off from Baghdad on the return trip via Damascus in January 1837, Rassam was deputed to accompany William Ainsworth, surgeon and geologist to the Expedition, on a subsidiary excursion into Kurdistan in search of coal[50] – vital of course for the operation of steamships. In the course of this Rassam returned to his native Mosul,[51] after an absence of ten years. His first intention, so he later told Palmer, was to try to preach the evangelical message to his people. But, so he said, he was overcome by shame when confronted by the aged Patriarch of the Chaldaean Church.[52] Then he went with Ainsworth through Anatolia to Istanbul,[53] and from there to England. On the way he visited the CMS mission station at Izmir, where he was disgusted, as he later said, by the heretical opinions of the two resident Swedish missionaries.[54]

In England Rassam went to Oxford, and there he visited, with a letter of introduction from Chesney, the High Church theologian who has been mentioned several times already: William Palmer (1811-79). This was the same cleric who later converted George Badger to 'Puseyism'. Palmer himself subsequently became a Roman Catholic, after unsuccessful attempts to join the Russian Orthodox Church.[55] Rassam clearly came very much under his influence, and adopted thenceforth a strongly anti-evangelical and anti-missionary position, denouncing the CMS as 'heretics, who destroy the Church & fancy Christianity to be a kind of spiritual Philosophy to which every man is to help himself out of the Bible as much as he can, or as much as he pleases'.⁵⁶ He also provided Palmer with some interesting details concerning the Malta establishment and the salaries of the missionaries there.⁵⁷ During the course of 1837 and early 1838 there was much contact between Rassam and Palmer, the former entering enthusiastically into the latter's Anglo-Catholic fervour: on one visit to Oxford, he even startled the customers of a local pub by haranguing them in broken English about the dangers of 'schism' (a matter of enduring interest to Oxonians). This was also the theme of a tract which he wrote in the form of a dialogue between a Christian and a Muslim, and which Palmer undertook to get published⁵⁸ (it never was, as far as can be ascertained).

The CMS, however, was as yet unaware of Rassam's change of heart, and it continued to entertain the expectation that he would return to his Arabic work in Malta. In November 1836, while Rassam was still in Iraq, Schlienz had looked forward to him resuming his work at the press, and had painted a rosy picture of a good, harmonious team of four Arabic editors and translators to undertake future projects: himself, Badger, Rassam and Fāris al-Shidyāq.⁵⁹ The following January, the CMS Secretary in London confirmed that Rassam was expected back,⁶⁰ and some time after his arrival in England contacted him with a proposal that he should indeed return and resume work at the Malta press, suggesting also that he participate in translating the Bible. According to Palmer, presumably based on what Rassam told him, the CMS 'applied to him to go & superintend their Press at Malta', but there is no hint of such a promotion in the CMS documents, and Schlienz was still in full control as Superintendent at that time. But however that may have been, Rassam, according to Palmer, then rounded on the Secretary and asked him 'whether the Society acted under the B[isho]ps & followed their orders', and on receiving a negative reply, berated the Society as 'bad people... & I will have nothing to do with you'.[61]

Instead, Rassam agreed to accompany Ainsworth on another expedition to north-east Iraq.[62] On the way, in the spring of 1838, he returned to Malta, presumably mainly to see his wife, from whom he had been separated for three years. Evidently to his surprise, she 'made no difficulty' on learning of his new travel plans.[63] As far as can be ascertained, he never returned to Malta. On completion of the Ainsworth expedition he was appointed by Lord Palmerston to be British Vice-Consul in Mosul, with a salary of £250 and permission to trade, and his wife eventually joined him there.[64] His appointment was said to be 'through the kind offices' of Viscount Ponsonby, British Ambassador in Istanbul.[65] He was the cousin of the late Sir Frederic Ponsonby, Governor of Malta, who had promised 'to provide something' for Rassam when he first left Malta in 1835.

In Mosul he assisted his brother-in-law George Badger in his mission to the Nestorians and in his campaign against the American Protestant missionaries, about whom he had already warned the Patriarch.[66] He revisited England in 1867/8, bringing a petition from bishops and clergy of the Nestorian Church of the East for help and support from the Church of England. Badger helped him to translate it and together they presented it to the Archbishop of Canterbury at Lambeth Palace, where it was favourably received.[67] Apart from this, he remained in Mosul for the rest of his life, trading for a while in partnership with the adventurer and businessman Henry Ross, whom he had previously known in Malta,[68] and later on his own. He was said after a while to have become the wealthiest man in Mosul; he lived in one of finest houses in the town[69] and he became renowned, together with his wife Matilda, for his assistance and hospitality to travellers and archaeologists, the best known of whom were Henry Layard and his own younger brother Hormuzd Rassam.[70] Later in life, he again took up translation work, as an amateur, rendering portions of the Old Testament into English, with glosses in Syriac and colloquial Arabic.[71] He died in Mosul in 1872.[72]

Rassam's was an interesting life, both in terms of the personal development of an adventurous Middle Easterner exposed to European influence, and in terms of the interaction between European travellers and one of their 'native' assistants who was far from being a mere servant or escort (although one British official called him 'the best nigger I ever saw'[73]). In parts it is rather difficult to unravel because of the web of half-truths and untruths which he himself had spun. This brief account has concentrated on the earlier part of the story, based on researches in the missionary archives and in Palmer's papers. A thorough study of the latter part, using the consular records and the accounts of 19th century travellers in northern Iraq, is a task which could usefully be undertaken.

## REFERENCES

1. Palmer, William, Notebook, Lambeth Palace MS 2817, pp.1-3. A brief summary of what Rassam told him about his background is also given in a letter from Palmer to his father in Palmer 1896, pp. 262-263.
2. Chesney 1868, p.555.
3. Pinches 1912, p.158.
4. CMS Archives: CM/065/20, Schlienz to Secs., 28.5.1832
5. Palmer, William, Notebook, Lambeth Palace MS 2817, pp.1-68. I am indebted to Dr J.F. Coakley of Harvard University for bringing this to my attention.

6. ibid., pp. 8-10.
7. Chesney 1868, p. 555.
8. CMS Archives: CM/O8/25, Rassam [to Secs.], 22.3.1834.
9. CMS Archives: CM/065/20, Schlienz to Secs., 28.5.1832.
10. CMS Archives: CM/O8/25, Rassam [to Secs.], 22.3.1834; Palmer, William, Notebook, Lambeth Palace MS 2817, pp. 10-11.
11. CMS Archives: CM/065/20, Schlienz to Secs., 28.5.1832.
12. Palmer, William, Notebook, Lambeth Palace MS 2817, p. 11.
13. CMS Archives: CM/O8/25, Rassam [to Secs.], 22.3.1834.
14. CMS Archives: CM/073/37, Kruse to Schlienz, 20.10.1829; CM/073/49, Kruse to Brenner, 2.7.1830; CM/O8/25, Rassam [to Secs.], 22.3.1834. Cf. Guest 1993, p. 71.
15. Perkins 1868, p. 307.
16. Ross 1902, p. viii.
17. Kubie 1965, p. 53; Mitford 1884, vol.1, p. 280.
18. CMS Archives: CM/073/49, Kruse to Brenner, 2.7.1830; CM/073/51, Kruse to Brenner, 30.10.1830; Palmer, William, Notebook, Lambeth Palace MS 2817, p. 12
19. CMS Archives: CM/073/49, Kruse to Brenner, 2.7.1830
20. ibid.
21. CMS Archives: CM/065/20, Schlienz to Secs., 28.5.1832. 'Fares' was the celebrated Lebanese writer and scholar Faris al-Shidyaq (1805/06-1887), who had worked in Malta for the CMS, 1826-28, and was to do so again, after Rassam's departure, between 1835 and 1842.
22. CMS Archives: CM/L2/252-253, Coates to Schlienz, 16.10.1833.
23. Perkins 1843, p. 52; id. 1868, p. 307; Kawerau 1958, p. 229; Hajjar 1970, p. 248. Hajjar mistakenly refers to it as 'une première ébauche de traduction de la Bible syriaque', evidently misreading Kawerau's 'Fibel' as 'Bibel'.
24. Wolff 1835, p. 523.
25. CMS Archives: CM/O8/25, Rassam [to Secs.], 22.3.1834; CM/L2/299, Coates to Schlienz, 24.5.1834
26. CMS Archives: CM/L2/330, Coates to Schlienz, 22.11.1834.
27. Palmer, William, Notebook, Lambeth Palace MS 2817, p. 19.
28 ibid.

29. CMS Archives: CM/L2/476, Coates to Schlienz, 26.1.1837; CM/O65/44, Schlienz to Secs., 3.2.1836. On Faris al-Shidyaq, *see* n.21 above.
30. *Missionary Register*, 1835, p. 404.
31. Palmer, William, Notebook, Lambeth Palace MS 2817, pp. 22-23.
32. ibid., p. 20.
33. CMS Archives: CM/O8/25, Rassam [to Secs.], 22.3.1834; CM/L2/299, Coates to Schlienz, 24.5.1834.
34. Chesney 1868, p. 165; Ainsworth 1888, vol. I, p. 3.
35. Chesney 1868, p. 166.
36. CMS Archives: CM/O18/22, Brenner to Coates, 9.4.1835.
37. CMS Archives: CM/O18/20, Brenner to Coates, 2.1.1835.
38. Guest 1993, p. 72.
39. Chesney 1868, p. 166.
40. Ainsworth 1888, vol. I, p. 4.
41. ibid., vol.II, p. 288.
42. Chesney 1868, p. 555.
43. Palmer, William. Notebook, Lambeth Palace MS 2817, p. 1.
44. Helfer 1878, p. 195.
45. A full account of the Euphrates Expedition is given in Guest 1992, from which the following very brief synopsis is mainly taken.
46. CMS Archives: CM/O8/27, Chesney to CMS, Port William, Euphrates, 27.9.1835.
47. CMS Archives: CM/O73/80, Badger to Schlienz, 28.11.1835
48. See above, n. 21.
49. CMS Archives: CM/L2/386, Coates to Schlienz, 1.1.1836
50. Ainsworth 1888, vol. II, p. 287.
51. ibid., pp. 313 sqq.
52. Palmer, William. Notebook, Lambeth Palace MS 2817, pp. 14-16.
53. Ainsworth 1888, vol.II, pp. 358-381.
54. Palmer, William, Notebook, Lambeth Palace MS 2817, pp. 21-22.
55. Palmer 1896, passim; Coakley 1992, pp. 21 sqq.
56. Palmer, William, Notebook, Lambeth Palace MS 2817, p. 20.
57. Id. Notebook, Lambeth Palace MS 2821, pp. 195-196.
58. Id. Notebook, Lambeth Palace MS 2817, pp. 73-158; Coakley 1992, pp. 21-24.
59. CMS Archives: CM/O65/45, Schlienz to Coates, 16.11.1836.

60. CMS Archives: CM/L2/476, Coates to Schlienz, 26.1.1837.
61. Palmer, William. Notebook, Lambeth Palace MS 2817, p. 68.
62. ibid., p. 67; Ainsworth 1842, vol. I, p. 2 .
63. Palmer, William, Notebook, Lambeth Palace MS 2817, p. 170.
64. Palmer, William, Notebook, Lambeth Palace MS 2819, pp.164 sqq., Chesney to Palmer, 28.12.1839.
65. Ainsworth 1842, vol. II, p. 334.
66. ibid., p. 251; Laurie 1853, p. 117; Coakley 1992, pp. 35-43
67. Coakley 1992, pp. 55-60.
68. Ross 1902, p. 13; Turner 2001, p. 118.
69. Turner 2001, p. 108, n.7.
70. Rassam 1897, p. 4; Turner 2001, pp. 128-129.
71. Pollington 1867, p. 355; Guest 1993, p. 120.
72. *The Times*, 9.7.1872, p. 1; Thielmann 1875, vol.11, p.116.
73. Letter from W. F. Williams (British representative on the Turco-Persian boundary commission) to Layard, 25.3.1859, cited in Turner 2001, p. 108, n.7

BIBLIOGRAPHY

*Published*

Ainsworth, W.F., *A personal narrative of the Euphrates Expedition*, London, 1888.

Ainsworth, W.F., *Travels and researches in Asia Minor, Mesopotamia, Chaldea, and Armenia*, London, 1842.

Chesney, F.R., *Narrative of the Euphrates Expedition*, London, 1868.

Coakley, J.F., *The Church of the East and the Church of England*, Oxford, 1992.

Guest, J.S., *The Euphrates Expedition*, London, 1992.

Guest, J.S., *Survival among the Kurds*, London, 1993.

Hajjar, J., *L'Europe et les destinées du Proche-Orient (1815-1848)*, Paris: Tournai pr., 1970.

Helfer, P., (Countess Nostitz), *Travels of Doctor and Madame Helfer in Syria, Mesopotamia, Burmah and other lands*, London, 1878.

Kawerau, Peter, *Amerika und die Orientalischen Kirchen*, Berlin 1958.

Kubie, N.J., *Road to Nineveh*, London, 1965.

Laurie, Thomas, *Dr Grant and the Mountain Nestorians*. Edinburgh, 1853.

Mitford, E.L., A *Land March from England to Ceylon Forty Years ago*, London, 1884.

Palmer, R. (Earl of Selbourne), *Memorials*, part I: *Family and Personal, 1766-1865*, Vol. I., London 1896.

Perkins, J., 'Attempt to interfere with the Nestorian Mission.' *Christian Work*, N.S.2 (1868), pp. 306-310.

Perkins, J., *A residence of eight years in Persia, among the Nestorian Christians; with notices of the Muhammedans*, Andover (USA) 1843.

P(inches), T.G., Rassam, Hormuzd. *Dictionary of national biography*, Second Supplement, Vol. III, London 1912, pp. 158-161.

Pollington, Viscount, *Half round the old world, being some account of a tour in Russia, the Caucasus, Persia and Turkey, 1865-66*, London, 1867.

Rassam, Hormuzd, *Asshur and the Land of Nimrod*, New York, 1897.

Ross, H.J., *Letters from the East 1837-1857*. ed. Janet Ross, London, 1902.

Thielmann, M. von, *Journey in the Caucasus, Persia, and Turkey in Asia*, London, 1875.

Turner, Geoffrey, Sennacherib's palace at Nineveh: the drawings of H.A.Churchill and the discoveries of H.J.Ross, *Iraq*, 63 (2001), pp. 107-138.

Wolff, J. *Researches and missionary labours ... during his travels between the years 1831 and 1834*, London (Malta pr.) 1835.

*Unpublished*

Church Missionary Society: Archives of the Mediterranean Mission, Birmingham University Library.

Palmer, William, Notebooks (1837-1850). MSS 2817-2821, Lambeth Palace Library, London.

Roper, G.J., *Arabic printing in Malta 1825-1845*. Unpublished thesis (PhD), University of Durham, 1988.

## Chapter Nine
# Mr and Mrs Smith in Greece, Egypt and the Levant

## Brenda E. Moon

John Shaw Smith was born in 1811 in London, but lived much of his life in Clonmult, County Cork, Ireland. He was the son of a linen draper of Gracechurch Street. He became a partner in his grandfather's firm of linen merchants, J & J Richardson & Company of Belfast. In 1839, at the age of 28, he married his cousin, Mary Louisa Richardson of Lisburn, County Antrim. He was a man of independent means and when their two children, Florence and Augustus, reached the ages of seven and nine, John and Mary Smith could afford to plan a grand tour of Europe and the Middle East. In 1849 Augustus was old enough to go off to boarding school, a French governess was found to take 'entire charge' of Florence, and having let their house in Regent's Park for three years, John and Mary set off for travels which would take them to France, Switzerland, Italy, Egypt and the Holy Land.

They were typical tourists of the time, but for one thing: John took with him a camera and calotype paper instead of the more usual sketch book and brushes, and produced some of the earliest photographs ever printed on paper of the countries which they visited. Thanks to the generous gift from his great grandsons, Peter and the late Basil Megaw, a set of prints made in about 1900 from John's negatives, and a typescript copy of the diary of the tour are now in Edinburgh University Library.[1] The original paper negatives and contemporary prints from them are in the Gersheim Collection, Humanities Research Center, Austin, Texas, and there is also a smaller collection of prints in the Palestine Exploration Fund archive. The original manuscript diary is in private hands. It is from the prints and typescript copy in Edinburgh that all the quotations and illustrations in this paper are taken. The section of the diary relating to the tour is mainly Mary's work, although there are occasional entries by John himself; it remains unpublished.

The calotype process was less than 20 years old when John purchased what he calls his 'camera and appendages' in February 1850. It was the invention of William Fox Talbot in the 1830s. Talbot discovered a method of producing negative images on coated paper, and by 1839 he was producing multiple positive prints from a single negative image, a process which has remained the basis of modern photography up to the present day. When John Shaw Smith bought his 'camera and appendages', the taking of 'views', as Mary calls them in her diary, was still an art requiring not only skill but patience. The average exposure was seven minutes in sunlight. John used a wooden camera with a 14 inch focal length lens and a three-quarter inch aperture. He never published his photographs, and remained an enthusiastic amateur all his life. It was only when some of his views were exhibited in the Victoria and Albert Museum in 1951, a century after they were taken, that they were recognised as masterpieces of Victorian photography.

After spending the summer of 1850 in Switzerland and much of the following year in Italy, John and Mary sailed on 14 October 1851 through the Straits of Messina at six in the morning and along the coast of Sicily. 'At length Etna made its appearance,' Mary wrote, 'a thin plume of smoke proceeding from its summit. As we approached close, the setting sun illuminated it so as to give the smoke a brilliant flame colour, which had a magical effect.'

The next morning they were in sight of Malta, where they landed 'in one of the brightly painted boats covered with an awning' and found the town of Valletta 'beautifully clean'. They were pleased to find sailors, guides and waiters at Dunsford's Hotel all speaking English. After a day of sightseeing they returned to the harbour and boarded the ship *Nile* for the Piraeus. Their fare from Naples to Malta was £5, and from Malta to Piraeus £7.10s. 0d.

After a stormy night they came in sight of Athens, with its dominant Acropolis, 'with the Parthenon proudly in the midst'. They were struck by the picturesque appearance of 'the Greek boatmen and others, all in the national costume ... the Phrygia cap, the jacket of various colours, white, red, blue and black covered with embroidery; belt or sash, and a very large, full petticoat of white cotton, the long gaiters, red, blue, white and black also embroidered.'

After four days of sightseeing they went in the evening of 19 October to the Grand Palace, where 'the object of all eyes' was 'some rope dancers; in the centre, the King and Queen, a little behind their attendants – the whole party on horseback'. Mary thought it 'very innocent of the Queen and Court, being so pleased with the rope dancers, but any change is agreeable here.' The next day they visited Eleusis, and admired the beautiful views on the way over the bay of Salamis. They alighted from their carriage to visit 'all that remains of the tombs of the Persians', and in the evening called on the Armenian wife of Mr Finley[2] to whom they had a letter of introduction.

The following day Mary wrote that 'John took six beautiful views, the best I think he has', while she went shopping with Madame Finley, who 'spoke such pretty broken English', and in the afternoon they sat near 'the glorious columns that remain of the Temple of Jupiter', where the view 'charmed us, modern Athens almost shut out... and we repeated those beautiful lines of Byron, "There is a Temple in ruins stands, &c &c" (Fig. 27). They spent two weeks in Greece. One evening they spent with Mrs Finley, 'about eight gentlemen and eight ladies, the latter sitting on one side of the room, and the former opposite. Mrs F. was very sweet and lively, dancing, singing and doing her utmost to amuse the company.' One Friday they spent on the Acropolis, where 'John took some good views.' Mary comments of the Parthenon that 'The Venetians and the Turks seem to have done their worst against it.'

They left Greece on 30 October in the *Tancredi*, 'a sad change to the *Nile*, so dirty; however there were few passengers so we got the ladies' cabin to ourselves.' After calling at Smyrna, where they found 'the people, Persian, Armenians, Jews, Turks and Greeks, the most curious part of the scene,' they reached the Dardanelles, and on 2 November they settled into the Hotel Byzantine at Pera, comfortably accommodated in two rooms with a 'charming view of the Golden Horn... close to a Turkish burying ground.'

They made several excursions by caique from Pera before visiting Istanbul, where they 'crossed a bridge in spite of the sentinels, one of whom', Mary writes, 'struck me a pretty smart blow on the head, but was quite agreeable with the others.' They visited mausoleums and mosques, and eventually reached the bazaar. 'Such a place I never saw! One tempting thing after another!' They purchased some slippers, and the next day too, 'after John had taken some views with the camera', they went back to the bazaar 'and had a long and amusing buying of a cap and chaplet, which took up two hours, pipes smoked and coffee. It is hard to know the price to give, as 400 piastres were asked, 175 taken.'

With permission they visited the old seraglio, where Mary soon discovered that some rooms had chairs exclusively for the Sultan's use, when 'having sat down, they immediately made me get up.' In the mosque of Santa Sophia they had another embarrassing experience when 'a fanatical-looking priest in a pulpit, declaiming with much vehemence to a crowd of men, women and children... suddenly commenced pointing at us and gesticulating, and we were forced to move on.' On 11 November John 'went out on his own with his camera and took some views in the afternoon,' Mary wrote: 'I did not go out again.' (Fig. 28). Two days later when they sailed for Smyrna, she was 'agreeably disappointed to find it quite quiet in the Sea of Marmora', but they had a rough crossing from Smyrna to Alexandria, and were glad when the lighthouse came into view.

They found Alexandria 'full, from the arrival of passengers from India. It can hardly be called an oriental city,' Mary wrote, 'but the moment you leave the town you find yourself completely in Egypt.' They took donkeys to visit catacombs three miles out and to make other excursions. John admired the mosque, but found the obelisk 'nothing remarkable'. In Alexandria they hired a dragoman, Hadji Bourni, for £6.10s.0d a month. 'He had good recommendations,' John said, 'though he was not much to look at.' Six days later they boarded the steamer for Cairo, proceeded through the canal, and in the evening changed boats and entered the Nile, now 'completely in Egypt'. On November 21 they reached 'the barrage, a gigantic undertaking, unfinished, and consisting of two vast dams to be thrown over the Damietta and Rosetta branches of the river... It seems doubtful if it is ever finished.' Their fare from Constantinople to Alexandria had been £10 each and the fare from Alexandria to Cairo was £3 each.

**Figure 27:** *The Temple of Jupiter, Athens. October 1851.*

**Figure 28:** *A street in Constantinople (Istanbul), November 1851.*

**Figure 29:** *A house in Cairo, November 1851.*

**Figure 30:** *Kom Ombo, January 1852.*

Figure 31: *Temple of Isis, Philae, December 1851.*

**Figure 32:** *The Great Temple, Abu Simbel, December 1851*

**Figure 33:** *From the walls of the Great Hall, Karnak, January 1852.*

Figure 34: *The Well of St. Stephen and travellers' apartments, Sinai, March 1852.*

**Figure 35:** *Petra, March 1852.*

**Figure 36:** *Baalbek, May 1852.*

In Cairo they settled into the Hotel d'Orient, and had a visit from a Dr Abbott, 'short, fat, dressed à la Turque, at first I thought him one. A dry little man, great antiquarian, fine museum of Egyptian antiquities, and married to an Armenian lady, very pretty, who lives in a latticed harem.' They lost no time in going down to Bulaq, and 'fixed on a boat in two days', but they stayed in Cairo for a week before boarding. Mary commented on the 'strange old lattices, the opposite houses nearly meeting', adding: 'John took two exquisite views' (Fig. 29). They explored the bazaars, Mary on horseback, but found the shops 'like cupboards, the people not nearly so agreeable as at Constantinople'. One day they visited the Tombs of the Caliphs, where the camera was again put into service. Mary mentioned seeing a funeral and also a wedding, 'the bride about eight or ten years old, covered up in a red veil and supported on each side by a tall, portly matron'.

They boarded their *dahabieh* on 1 December, and were soon 'extremely busy in arranging our goods and chattels, putting up shelves, [and] adorning our cabin. Begin to wear a more comfortable appearance.' Then began their journey upstream. One night they anchored next to 'a boat of Americans. John paid them a visit, one being a photographist!!!' They referred to it as 'the rival boat' as they repeatedly passed each other on the journey. Another boat anchored close, carrying a Captain Dent and a Mr Shipley, nephew of Bishop Heber. John went on board, and must have reported favourably to Mary, who wrote approvingly that 'they have [her] very nice and clean and well arranged. Afterwards they came to see ours, with which they expressed themselves much delighted.' She commented on a steamer which passed them, 'looking very much out of place in this scene'. On 6 December she was still home-making: 'I put up my new curtains and drapery which look very nice indeed and improve the cabin much.'

Life on board was not, however, without its problems.

'The crew and reis [were] in a state of mutiny on the subject of carrying three sails... John had to get up in the night because of the sail. We are going to take strong measures on the subject. No *backsheesh*.' Two days later there was a grand fight about the third sail, and the boat ran aground. John and Mary both fell about, but were uninjured. They stood their ground with the crew and won the argument.

By mid-December it had turned 'very cold indeed', but they enjoyed a variety of shore excursions, one to a Coptic monastery, where they found the people very agreeable. By now they had become used to running aground, generally 'three or four times a day'. On 15 December they passed 'a handsome pigeon-house and palm grove; [it] would make a charming picture,' wrote Mary: 'the most striking objects in villages are the pigeon houses... like square towers... John took two pretty views of a sheikh's tomb and trees and some pigeon-houses... very characteristic.'

Two days later they landed to visit the Temple of Luxor. '[A]nd what shall I say?' she wrote: 'Why, I was disappointed in it!!! It was less grand than I expected, but it is hard to judge of it half buried as it is and all built within and without with those horrid mud hovels which stop up the view wherever you turn.' Of the colonnades she wrote: 'Such a confused mass altogether that I could not imagine the whole,' but when they were again afloat she conceded, 'the Temple of Luxor did indeed look most imposing as we glided by.' While she was writing lyrically that evening, as often, about the sunset, a violent shock as they hit a sandbank brought her to earth and 'scattered our library all over the room'.

When they moored at Esneh another boat came up bearing a Mr and Mrs Giles and a Mr Yatman. 'Quite pleasant to see a lady again,' commented Mary. They found that many of the women of Esneh were going about unveiled, and she described the town as 'the spot of banishment for the dancing women of Cairo'. But she admired Kom Ombo (Fig. 30) and the bas-reliefs on the roof of the temple at Esneh, and the beautiful state of preservation of the walls, 'as fresh as if newly cut from the chisel. Some *morceaux* I should like to carry away. The part excavated is merely the portico, the remainder... being still buried.' On their return to the boat John prepared his views, while she went out with Mrs Giles.

They spent two days at Aswan, rowing across to Elephantine one day, visiting the unfinished obelisk on another, and riding along the river bank to view the rocks and rapids from above, before venturing up the cataracts. Mary found the ascent the next day 'most exciting', but they were glad to 'float calmly into the beautiful basin above which is the first view of the glorious temple of Philae ... Our wonder and delight was extreme... It is rarely one has the satisfaction to enjoy in silence and solitude the mighty monuments left by the ancient kings... John took four beautiful views of the temples' (Fig. 31).

There are occasional domestic details in Mary's diary: 'I had a washing and ironing,' she wrote on 23 December, and one Sunday she wrote: 'Thought it was Saturday and actually mended my stays. How grieved I am!' The Victorian Sunday was most certainly recognised on the Nile.

Beyond Philae they began to see 'very wild-looking black people – almost naked, many being entirely so, and very disgusting. The change in the inhabitants since we passed the cataracts is astonishing.' But she did not find all black people disgusting: at Dendur she describes some of the crowd of women who greeted them as 'very handsome', and at Tell el 'Amarna which they visited on the way down they 'saw such a lovely little black child near the Temple. I longed to buy it but its mother would not part with it.'

On 29 December they reached Abu Simbel, and explored the temples: 'first the small one... the walls covered with hieroglyphics or ciphers, very interesting... and from that I was dragged up a hill of sand to the grand Temple... In the countenances [of the colossi], vast as they are, is an expression of grand benevolent beauty which is most imposing and wonderful... John took four good views, one in particular of the colossal Rameses, is admirable' (Fig. 32).

On New Year's Eve they rode on donkeys to the second cataract and climbed the rock of Abusir. Mr and Mrs Giles had already arrived and were 'romantically cutting their names on the rock. I saw Belzoni's, which I had serious thoughts of taking with us, but Mr Giles said it would be atrocious, so [he] being a clergyman, I suppose, I listened to him; however I think he had an itching for it himself, and was jealous lest I should carry it off.' Mary wanted to ride on further to Semna, 'and I should have carried the day, but alas there is no tent to be had where a lady could dress reputably in the morning.'

On New Year's Day they began the journey downstream in company with Mr Giles ('profound and hieroglyphical') and Mrs Giles ('facetious and loquacious as ever'). When they reached Philae 'John took some beautiful views... from the opposite island of Biggeh [Biga]', and they returned to the Temple by moonlight, but 'we both got horribly melancholy and hurried away. Great disturbance of Americans.'

John took more views at Edfu, and of course at Thebes (Fig. 33), where they explored both temples and tombs, but Mary recoiled from the 'very bad smell, and it is disgusting and horrible to see the remains lying about and trodden under foot in all directions.' After that she left most of the tomb-visiting to John.

At Karnak John's first attempts at photography failed, 'why – we cannot imagine'. Mary writes, cryptically, 'It seems the reason [is that] the Sheikhs don't like their tombs to be taken, and so the views don't succeed.' They found, as so many others did, that once they had turned northward they began to long for Cairo and news from home. 'How anxious and nervous I feel,' Mary wrote on 10 February while John was out collecting their post in Cairo, and then, 'John has come. Good news from the darlings and from home. How relieved I am, but alas, we are still very dull and sad.'

Their boat when they left it on 13 February 'looked very forlorn', but their spirits revived with visits to Saqqara, Memphis and Giza. Of course they ascended the pyramid of Cheops, but 'we did not add our names to the numbers that deface the stones.' It was 'a capital wind-up to our voyage on the Nile.'

But they had more adventures ahead. With a new dragoman, Abdalla, after a row with Hadji Bourni, they rode out on 23 February for Sinai and a month in tents. 'It is necessary to plan a little,' wrote Mary, 'as our tent is small, and [we] have two good beds, but we have now found a convenient place for each article.' On 26 February they reached Suez, and stayed at the Hotel of the Transit Company, hoping to get a boat to Ayoum Mousa, but the wind was so high that no boats were sailing, so they rode round by the coast. 'We had hoped to have found all things ready for us – tent and dinner after a long day – and very fatiguing from the high wind, but it was very late before we got settled.' Mary was enchanted by the scenery as they rode through a succession of defiles and small valleys. On 2 March the temperature reached '105%' [sic]. 'Our encampment with the men and the camels make such a pretty picture in the midst of these wild mountains.' When their route lay along the sea shore, they bathed in the sea.

On 6 March they were again in the desert and, riding up 'to an Arab encampment, saw a number of dirty, veiled women with a profusion of necklaces. Our sheikh, who had gone on ahead the day before to see his family,' reappeared with his mother. 'She showed us her face. Fearfully ugly. It was wise of her to wear a veil.'

The next day they began the ascent to the Convent on Sinai, and reached it during the day (Fig. 34). They were accommodated in the apartments for strangers, and explored the library, the Chapel of St Katherine, the Chapel of the Burning Bush. On 7 March they ascended to the Chapel of Elijah and the two chapels on the summit of the mountain. 'John went to visit the rock struck by Moses... I returned very tired to the Convent, but delighted that I had accomplished this most interesting pilgrimage.'

They left Sinai for Aqaba on 10 March. 'There was a great noise and confusion all the morning... the monks, one after the other, running out to ask *backsheesh*... and one of them most pertinaciously wanted our table... Much regretted we had not remained in our tents as these little annoyances destroy much of the deep feeling with which the spot cannot but be visited.'

On 13 March they reached the sea and came to 'an exquisite spot secluded and tranquil as if no chance visitors like ourselves had ever trodden its shores.' There they camped, and 'John retired to another little bay and suffered severely from the treacherous beauty of the coral reef.'

Soon they found themselves in the territory of 'Sheikh Hussein' whom Mary called 'that bug-bear to travellers'. He called on them just as she was preparing to wash and dress. 'We called for pipes and coffee, and flattered him to a ridiculous manner and declared at the same time our firm resolve to go to Petra, which he seemed to say was impossible, and after some time he went away.' However, on 17 March he returned, and said that he was ready to take them to Petra. They chose their camels and began the journey – a large caravan, four parties together, with three sheikhs. Mary was ecstatic at the scenery of Mount Hor, and delighted when they found their tent 'pitched in a delicious grassy spot... in the middle of the ruined city, one of the most wonderful in the world, and a long-wished for object of our [travels].' John went out to take photographs (Fig. 35), while Mary went for a ramble in the ruins, but 'had not gone far when a guard overtook me, and remained with me for the rest of the time', until she returned by a short route to where John was 'finishing his views', and found 'Abdalla in a great fright at my escapade'.

Two days later they struck camp and climbed the pass by which they had come, and left behind 'this wonderful city of the rocks for ever!! It far surpassed even our highly raised expectations.'

On 28 March they reached Judaea, and were 'captured by the quarantine guards'. They entered the quarantine house and waited for horses and mules to arrive. Three days later 'we ride out of prison'. By 3 April they had arrived in Jerusalem, and visited the Church of the Holy Sepulchre on Palm Sunday. They lodged in a house in the Via Dolorosa; 'desolation and dirt reigned in its interior,' wrote Mary, 'the only time in all our travels that Abdalla did badly for us.'

During their time in Judaea they followed a pilgrimage to the Jordan, bathed in the Dead Sea, and visited Bethlehem, as well as all the sacred sites in Jerusalem itself, before turning north to Tiberias on 27 April. From there they visited Mount Carmel and Nazareth, then on May 7 they arrived in Damascus, 'this charming city... so quiet and tranquil'.

At last, on May 21, they reached Baalbek, the last goal of their Levantine wanderings, and had their tents pitched in the 'vast enclosure'. The next morning John 'went out to take views' (Fig. 36). After some days exploring the ruins and underground corridors they left on 26 May and encamped for the last time 'in the tent where we have spent many a happy hour. We feel rather dismal, and sing for a long time in the twilight.'

At Beirut they said goodbye to Abdalla. He brought Mary a bouquet and a bunch of bananas, 'kind and attentive to the last, doing all in his power.' It was time for home.

# REFERENCES

The permission of Dr. David Shaw Smith to quote from the diary of the tour, of which he owns the original manuscript, is gratefully acknowledged.

1. Edinburgh University Library, MSS E99.41. The permission of the Librarian for quotation and reproduction is gratefully acknowledged.
2. George Finlay, 1799-1875, philhellene and historian of medieval Greece.

## Chapter Ten
# Robert Murdoch Smith and the Mausoleum: Excavations at Halicarnassus (Bodrum) 1856-59

# Jennifer Scarce

Scots have a long history of involvement in the Near and Middle East, quickly taking advantage of the opportunities which an increased British presence gave them to pursue careers as diplomats, soldiers, merchants, scientists, artists, archaeologists and explorers.[1] The career of Robert Murdoch Smith (1835-1900) was impressively versatile: he moved capably between political, military, technical and archaeological roles which took him to Turkey, Libya and Persia.[2] His activities at Halicarnassus (1856-59) both employed his technical skills as a young officer in the Royal Engineers and introduced him to the classical archaeology of Asia Minor and the contemporary culture of Ottoman Turkey (Fig. 37). His involvement here should be properly understood in the context of the changing nature of the European obsession with the civic and architectural ruins of classical Greece and Rome, many of whose most impressive examples are located in modern Turkey.³

A particularly British aspect of this obsession in the 17th and 18th centuries was a fascination with classical sculpture which provided the 'picturesque ruins' essential for the painstakingly contrived naturalism of English formal parks and gardens. Thomas Roe, ambassador to Constantinople from 1621 to 1628, had helped the Earl of Arundel and the Duke of Buckingham to acquire figures from Asia Minor to decorate their homes and gardens.⁴ During the 18th century Horace Walpole toured the great houses of England and noted this fashion. Increasingly improved conditions of travel also made the sites of classical antiquity more accessible and familiar. Here the grand tours of Greece and Italy completed the education of young aristocrats, while wealthy patrons and their entourages of scholars, architects and artists explored Turkey, the Levant and Egypt, frequently recording their visits with portraits of themselves in glamorous Ottoman dress against suitable classical backgrounds such as the Parthenon.⁵ The Earl of Sandwich, for example, began his lifelong interest initially on his grand tour of 1738 when he travelled to Constantinople and then through Turkey to Egypt. Later he joined the Society of Dilettanti, founded in 1735, the most active of the many antiquarian societies whose study of surviving monuments ensured a continuing enthusiasm for classical art and architecture.

The high standard of recording which had developed from the activities of the 18th century continued in the 19th, supported by a parallel interest in archaeology. Apart from the wealthy private scholars, institutions such as the British Museum were building up collections of classical objects. During expeditions to Lycia in the 1830s and 1840s, Sir Charles Fellows acquired a large collection of sculptures, whose transportation to the British Museum from the Xanthus river was organised by the Royal Navy under the command of a Scot, Lieutenant C.E.J. Ewart. These and other collections were systematically arranged in galleries, with the aim of showing the public the origins and evolution of classical art.[6] Contemporary acquisitions of Assyrian and Egyptian sculpted figures and reliefs offered material for scientific comparison. The Museum's interests in archaeology provided the ideal background for the work of Charles Newton (1816-94), who introduced Murdoch Smith to the Mausoleum.

Charles Newton entered the Department of Antiquities of the British Museum on 10 March 1840 as an Assistant Keeper, a post he was to hold until 1851. He was recruited to meet the Museum's need for qualified staff to cope with the demands of an expanding programme of acquisition, exhibition and cataloguing. He had developed a keen interest in the discipline of archaeology as an undergraduate at Oxford, but spent most of his museum career labelling coins and editing publications of the Department's collections of sculptures and vases. His only chance of experiencing classical material at first hand came in 1848 when he was granted three months special leave to visit sites in Athens, Naples and Rome.

While he had appreciated the training and discipline which the Museum had given him, he realised that his opportunities to pursue archaeological fieldwork would be limited. He therefore resigned in 1851 to take up a post in the Foreign Service as Vice-Consul at Mytilene, where he served until 1860. During these years he kept his links with the British Museum and was able to combine his official duties with expeditions to neighbouring Greek islands and the Turkish mainland. He recorded and collected inscriptions, sculptures and vases for the Museum,[7] supported by modest funds from the Department of Antiquities and official permits and finance obtained through the influence of Lord Stratford de Redcliffe, British Ambassador to Constantinople.[8] His real chance of making a major contribution to classical archaeology came in 1855 when he visited the small picturesque Turkish coastal town of Bodrum, once Halicarnassus (Fig. 38), now a major tourist resort. Halicarnassus was a Dorian Greek colony of the 9th century BC in the state of Caria, and the site of the Mausoleum, one of the Seven Wonders of the Ancient World,[9] which became the generic name for any grandiose funeral monument.

Halicarnassus had an eventful history, shifting allegiance between competing leagues of Greek states and colonies before coming, in the 5th century BC, under the control of the Achaemenid Persians, who allowed local rulers to govern the province of Caria. The most renowned local ruler was Mausolus who ruled from 377 to 353 BC (Fig. 39). Mausolus and his family were deeply influenced by Greek principles of town planning and architecture – principles seen in their development of Halicarnassus as their capital. New walls, temples, a palace, theatre, market place and other public buildings were constructed, but the most prominent and magnificent edifice was the marble decorated tomb – the Mausoleum – which was probably begun in about 370 BC and completed in about 350 BC after the deaths of Mausolus and his wife Artemisia. The Mausoleum was indeed spectacular: it soon became famous throughout the ancient world and was praised by classical authors. The account of Pliny the Elder written in about 75 AD gives the clearest description.[10] Rectangular in plan, the Mausoleum was 140 feet high and consisted of three parts – a base, a colonnade of 36 columns and a stepped pyramid ascending to a platform crowned by a marble chariot drawn by four horses. The Greek art and architecture which Mausolus so admired was reflected in his tomb; the colonnade used the graceful Ionic order, while the sculpted decoration employed, according to Pliny, the finest contemporary sculptors – Scopas, Bryaxis, Timotheus and Leochares – who worked on the reliefs on the four sides, and Pytheos who was responsible for the chariot. This remarkable concentration of classical sculpture was not, however, to survive intact; today there is virtually nothing left of the Mausoleum on its site north-west of Bodrum.

After Mausolus' death Halicarnassus continued its eventful history. It fell to Queen Ada, an ally of Alexander the Great in the 4th century BC and after her death in turn to the Ptolemies of Egypt, the Seleucids of Syria and, in 129 BC, to the Romans. The Arabs sacked it in 654-655 AD, the Menteseoglu Turks took it in the mid-13th century, the Knights of St John of Malta seized it in 1402 and finally it came under Ottoman rule in the 16th century. Surprisingly the Mausoleum survived many of these changes of ruling power. Bishop Eustathius briefly noted it in the 12th century but the upper part seems to have fallen down during an earthquake in the 18th century.

The Knights were responsible for its real destruction after they decided in 1494 to extend the massive Castle of St Peter which they had begun in 1402. They pillaged the Mausoleum for building materials, using blocks of green volcanic stone and marble sculpted friezes to construct the castle walls. Between 1505 and 1507 a dozen slabs of the frieze showing a battle between the Greeks and the Amazons, a single block from a battle between Lapiths and Centaurs and foreparts of four statues of standing lions and a running leopard, were built into the Castle. Other sculptures were not so fortunate as they were broken into small pieces and burned to make lime mortar.[11] This work of destruction continued until 1522. Travellers, however, continued to be aware of the Mausoleum. A French traveller, Claude Guichard, published in 1581 an account of the discovery and pillaging of the Mausoleum in 1522;[12] Richard Dalton published drawings of the friezes in the Castle walls which he had seen in 1749. The friezes only received active attention in the 19th century stimulated by the negotiations of Lord Stratford de Redcliffe. He had obtained in 1844 a firman from the Ottoman government to send Charles Somers to Bodrum to look inside the Castle. He saw there the classical friezes and sculptures embedded in the walls. In 1846 Lord Stratford obtained permission to remove twelve reliefs of the Amazon frieze and ship them to London. They arrived in the British Museum between 5 and 9 September and were put on display under the supervision of Charles Newton.

Newton, who had seen the white marble lions still embedded in the walls of the Castle during his first visit to Bodrum in 1855, was inspired by these acquisitions, and was immediately able to put them into the context of the Mausoleum. He was determined to undertake archaeological fieldwork to locate the site of the Mausoleum. After another visit in 1856, where he managed to excavate many terracotta figurines in the field of a local Turk,[13] he returned to the British Museum with a proposal supported by charts and plans of the Bodrum area prepared for a fieldwork campaign by Thomas Spratt of the survey branch of the Royal Navy. Newton's application was successful. He obtained the necessary firman to excavate from the Ottoman government, and then applied to the Foreign Secretary Lord Clarendon for £2,000, plus equipment, a ship of war for six months to ensure a supply line of food and water, and a core of technical staff to be recruited from the Royal Engineers. His requests were granted and he was given the *Gorgon*, a steam corvette with a crew of 150 men commanded by Captain Towsey. This is where Lieutenant Robert Murdoch Smith enters the story, for he was one of Newton's contingent of Royal Engineers. Also with Newton were Corporal William Jenkins as senior NCO, Corporal B. Spackman as photographer (the first to accompany any archaeological expedition), and two resourceful Lance-Corporals, Patrick and Francis Nellis, who served respectively as smith and mason. They had been born in Corfu and spoke Greek and some Turkish.

The expedition opened a new world of adventure and opportunity to Murdoch Smith, who was perhaps typical of the young Scots of modest respectable families who found employment abroad. He was born on 18 August 1835 in Kilmarnock to a doctor, Hugh Smith, and his wife Jean Murdoch, a farmer's daughter. He was educated at Kilmarnock Academy, where he gained a thorough knowledge of the Greek and Latin classics, and then went on to Glasgow University. Here he combined arts and sciences, continuing his classical education and studying chemistry, and moral and natural philosophy. It is probable that his required reading would have included volumes of George Grote's *History of Greece* (completed in 12 volumes between 1846 and 1856), which was the standard text of 19th century courses in ancient history.

His extracurricular studies were enterprising and varied. He was an excellent linguist, acquiring fluent French and German, a working knowledge of Italian, and some Arabic. He graduated with a sound education and a need to find a career, as he could not depend on a private income. Army service was possible, but here again he was limited as he could not afford to purchase a commission in a fashionable London regiment. His options were the Indian Army and the Royal Engineers. The Engineers valued professional and technical qualifications above social and financial connections, and Murdoch Smith chose them, entering the first open competition for commissions in the Ordnance Corps on 1 August 1855. He came first of 380 candidates in the qualifying examinations. He was then gazetted to a Lieutenant's commission on 24 September 1855 and stationed at the Royal Engineers establishment at Chatham on 18 October 1855. After such a promising start Murdoch Smith might well have enjoyed a good army career; his talents would have ensured steady promotion to high rank. Royal Engineers, however, travelled widely and adventurously. They were in the Crimea and in China; they

**Figure 37:** *Lieutenant Robert Murdoch Smith R.E. in 1856 (from a colour photograph by Alexander Stanesby).*

**Figure 38:** *Bay of Bodrum with the Castle of St Peter (author's photograph).*

**Figure 39 (left):** *Marble statue of Mausolus c. 350 BC (British Museum, Greek & Roman Department, 1000, author's photograph).*

**Figure 40 (right):** *Marble statue of Artemesia c. 350 BC (British Museum, Greek & Roman Department, 1001, author's photograph).*

**Figure 41:** *Marble lion from the Mausoleum c. 350 BC (British Museum, Greek & Roman Department, 1075, author's photograph).*

**Figure 42:** *Sir Robert Murdoch Smith in 1898 (from a photograph by R.S. Webster, Edinburgh).*

built bridges, roads, established water supplies, and repaired walls and houses. It was natural, therefore, that their technical skills should be required in the service of the developing discipline of field archaeology. So it was that the 21 year old Murdoch Smith sailed on the *Gorgon* out of Spithead in 1856 and arrived in Smyrna on 11 November. He was to devote his services to the British Museum's classical expeditions in Turkey until 1859.

Excavations at Bodrum began soon after Murdoch Smith's arrival there on 24 November 1856, in the field where Newton had dug in 1855. As this only yielded more terracotta figurines and lamps, Newton decided to explore other areas. This he did profitably at intervals throughout December, finding, for example, the remains of a large Roman villa with a mosaic floor which Corporal Spackman photographed.[14] Exploration of this complex and extensive site was long and difficult. It involved patience and strategic decisions about locating the best places for trenches and handling of finds, and he was involved in tricky negotiations with local Turks.[15] Apart from Newton's official publications the story is told in the diaries and letters of Captain Towsey and particularly of Murdoch Smith. His letters written between 1856 and 1859 to General Sir John Burgoyne, his commanding officer in the Royal Engineers, are an informative and lively record of progress and reveal also how essential a role he played in major discoveries. He clearly understood that :

> The object of the expedition, as explained to me by Mr Newton, is to excavate at places where Hellenic traces have been found, and discover as much as possible of the ancient city. The principal points are the Mausoleum, the temple of Mars, the Temple of Venus, the Palace of Mausolus, and the Rock Sepulchres (letter 26 November 1856).[16]

In the same letter he clearly described the visible remains of the ancient site and noted Newton's methods of excavation which consisted of directing his men to dig trenches in all directions, tossing the earth behind them as they advanced. Eventually the results were more rewarding. Murdoch Smith's work was groundbreaking: he discovered the site of the Mausoleum. In his letter of 1 January 1857 he stated:

> On the 27th we commenced digging in another place. As I happened to point it out as the probable site of the Mausoleum it goes by the name of Smith's Platform. It lies on the rising ground a little to the north of the Pasha's palace, which you will see marked on the chart. From the number and appearance of the marble fragments, and its central position, I was always of the opinion that the Mausoleum must have stood in the neighbourhood and since we have commenced our excavations there, every day brings additional proof of its being the actual site.[17]

He correctly matched architectural fragments from the site with those embedded in the Castle walls and identified pieces of marble Ionic columns and remains of the cornice and sculpted frieze. His letter of 3 February 1857 revealed finds of more Ionic columns, pieces of moulding and sculptures from the architrave and cornice which further confirmed the identification of the Mausoleum. He also realised that:

> It would appear that the Mausoleum had at least two friezes – one, part of which is in the British Museum, and another of much higher relief, and probably considerably larger.[18]

More spectacular finds were to follow, as reported in Murdoch Smith's letters of 2 April and 2 May 1857. After continued excavation of the Mausoleum's foundations he recorded:

> We have lately found several fine pieces of sculpture, chiefly parts of lions and fragments of friezes. The pieces of lions have always, with one exception, been the hindquarters, the heads and foreparts having probably been taken away by the Knights, as in the case of those that were in the castle wall... The lions in the castle were all taken without accident.
>
> This morning we discovered a colossal statue of a female figure sitting in a chair... A few days ago we found the body of another large female figure. They are both remarkably fine. As I hope to send photographs of them next time I write, I need not describe them more minutely. One of them is probably Artemisia, the wife of Mausolus, and the builder of the Mausoleum (2 April 1857)[19] (Fig. 40).

A month later he reported:

> Our excavations have been continued last month at the Mausoleum with even more than our former success. They have been carried on principally on the north side, where we found the foundations of a marble wall, that seems from its appearance and dimensions to have been one of the walls of the *peribolus*... (2 May 1857).[20]

More sculptures later emerged:

> The first of them was a lion, almost perfect, part of his legs only being wanting. It corresponds exactly with those formerly found, and with the others from the castle wall. (Fig. 41).
>
> In the same place as these statues we have just found the hinder half of a very large horse... Mr Newton says it is the

largest piece of Greek sculpture that has ever been discovered. Pliny, in describing the Mausoleum, says that the pyramid, which consisted of 24 steps, was surmounted by a chariot with four horses sculptured in marble by Pythis... As the horse we found has no rider, it must evidently be one of the four of the *quadriga*[21] (2 May 1857).

Murdoch Smith then presented his results in a technical report of 1 June 1857 which formed the basis of the architectural restorations prepared for Newton's official publication of the expedition to the Mausoleum.[22]

After these discoveries and the departure of the *Gorgon* on 24 June 1857, laden with cases of sculptures for the British Museum, Murdoch Smith's role at Halicarnassus was effectively over. Newton did continue to explore the Mausoleum, finding more sculptures, including a colossal ram. He decided, however, to move down to the ancient city of Cnidus in December 1857, inspired by the reports of its classical remains prepared by members of the Dilettanti Society after their visit in 1812. He left Murdoch Smith to complete operations at Halicarnassus, which he did despite shortage of manpower and severe winter weather. Murdoch Smith was to continue working in Turkey until 1859. By March 1858 he had moved to Cnidus, where Newton required his technical expertise. Excavations there, concentrating on the Sanctuary of Demeter, had yielded rich finds, including a fine seated figure of the goddess, inscriptions, terracotta figurines and lamps and glass vessels. The real prize, however, was discovered on a rock some distance from the main site – a colossal marble lion six feet tall and about nine tons in weight. Murdoch Smith spent a month raising the lion, lowering it into a packing case and transferring it to the ship *Supply*. His ingenious solution to the problems involved is described in enthusiastic and technical detail in his letter from Cnidus of 1 July 1858.[23]

After a well-deserved period of home leave during the winter of 1858-59, Murdoch Smith returned to Cnidus in March 1859, where again he supervised the final stages of excavation. He left Turkey on the *Supply* in late May for Malta, where he was employed on garrison duty until November 1860.

The archaeological campaign of 1856-57 to discover the Mausoleum of Halicarnassus was to affect the future of the monument itself, and of the two main personalities involved in the expedition, Charles Newton and Robert Murdoch Smith. The glories of the Mausoleum's sculptures were revealed to a wide public. The sculptures arrived on the *Gorgon* at Woolwich on 22 July 1857 and were transported to the British Museum, which had to arrange storage, conservation and display. It is much to the credit of the Museum that the most significant sculptures were assembled and exhibited in a temporary arrangement of sheds within the main external colonnade. The site itself remained an intriguing enigma. It required considerable reserves of imagination from the visitor, as so little was visible of the remains of the Mausoleum and the once-thriving ancient city. Between 1966 and 1977, teams of excavators led by Professor Kristian Jeppesson of Aarhus University managed to clear the site of debris and scattered fragments. This enabled them to reconstruct a more accurate plan of the tomb.[24] The Graeco-Roman theatre, which has an estimated seating capacity of 13,000 and is situated north of the Mausoleum, has also been excavated and restored. It offers a good overview of the site, the castle and the bay of Bodrum.

The Mausoleum's reputation as one of the Seven Wonders of the Ancient World enshrined it in Bloomsbury's memory well before the British Museum acquired its remains. The parish church of St George's, consecrated on 28 January 1730 by Edmund Gibson, Bishop of London, to serve the parish of Bloomsbury, was designed and built by

Nicholas Hawksmoor between 1716 and 1731. He drew on a repertoire of classically inspired architectural form and decoration to construct a handsome church with a colonnaded portico. Its tower was inspired by Pliny's description of the Mausoleum, but a statue of George I replaced the marble chariot.

Charles Newton's career as a classical archaeologist was confirmed by his work at Halicarnassus. On his return to London he concentrated on plans for the publication of the site, and studied the sculptures with a view to their permanent display. He was eventually able to realise his plans when, on 17 June 1861, he was appointed Keeper of the newly created Department of Greek and Roman Antiquities in the British Museum, a post which he was to hold until 1886.[25] The major achievement of his Keepership was the arrangement of the Museum's large collection of classical friezes and sculptures. His solution was to divide them into stylistic groups arranged chronologically in order to trace the evolution of Greek sculpture from its archaic beginnings to its accomplished maturity. In this scheme the Mausoleum's sculptures and friezes were rescued from temporary display and permanently exhibited in a special gallery. Here the figures of Mausolus and Artemisia were given prominent positions, while the friezes showing battles between Greeks and Amazons lined the walls.

Halicarnassus brought promotion for Charles Newton and stimulated Murdoch Smith to find new opportunities outside his military duties. Garrison life at Malta was not demanding enough to make the best use of Murdoch Smith's talents, and his application in 1859 to serve in China was rejected because all the troops required there had been sent from India. His experiences at the Mausoleum, supported by his sound technical and classical qualifications, had sharpened his appetite for archaeological field research. He was aware that the Libyan site of Cyrene, capital of a Greek colony and then of the

Roman province of Cyrenaica, was relatively near, and had never been adequately surveyed and excavated. He obtained leave of absence and, in November 1860, with Lieutenant E.A. Porcher R.N., departed from Malta to excavate Cyrene at his own expense. By the autumn of 1861 they had surveyed and planned the site, excavated the Temples of Apollo, Venus and Bacchus, and shipped cases of sculptures to the British Museum. Their account of the excavations was prepared while Murdoch Smith was employed at the War Office from 1861 to 1863 and published in 1864.[26] As well as deepening his knowledge of classical antiquity and exercising his technical skills, archaeology had given Murdoch Smith opportunities to initiate and direct significant projects and to develop an understanding of contemporary life in Turkey and Libya.

This experience was to equip him for the next and longest stage of his career when, in 1863, he answered an advertisement in *The Times* for Royal Engineer officers to supervise the work of constructing a land telegraph line between London and India, via Persia. It was the ideal opportunity for him. He achieved distinction as Director of the Persian Telegraph Department (1865-88). During his years in Persia he found time in a busy professional life to maintain his enthusiasm for archaeological and historical research and to investigate Persian Islamic art. As with his work at Halicarnassus, museums benefited from his knowledge and expertise. He built up collections of Persian art for the Victoria and Albert Museum and, after his retirement from the Telegraph Department, for the Edinburgh Museum of Science and Art (now the National Museums of Scotland) where he served as Director until his death in 1900[27] (Fig. 42).

His achievements laid the foundation of a proper study of traditional and contemporary Persian art, which was shown to the public for the first time in a great exhibition held at the Victoria and Albert Museum in 1876. Here the results of his systematic methods of collecting and documenting related collections of ceramics, metalwork, lacquer, textiles and carpets ranging in date from the 16th to 19th centuries, were summarised in a catalogue and handbook which is still an essential reference work. The Edinburgh Museum of Science and Art similarly benefited from Murdoch Smith's knowledge and experience of acquisition, exhibition and publication of Persian art. He was also enlightened and curious enough to take an active interest in contemporary Persian arts and crafts, acquiring the work of a tilemaker named Ali Muhammad Isfahani, and encouraging him to write a treatise describing his materials and methods. He well understood the role of the modern object in the accurate documentation of the past.

## REFERENCES

1. For example, George Strachan of the Mearns, who in the early 17th century was one of the best-qualified employees of the East India Company in Persia. He had an excellent knowledge of Arabic, Persian and Turkish and collected books and manuscripts written in these languages.
2. Dickson, W.K., *Life of Major-General Sir Robert Murdoch Smith*. London and Edinburgh: Blackwood 1901.
3. For example, Aphrodisias, Aspendos, Ephesus, Perge, Priene, Side.
4. The Arundel Collection is now in the Ashmolean Museum, Oxford.
5. Examples of British travellers in Ottoman dress include Lord Byron, Thomas Hope, David Roberts, Robert Wilson.
6. See Jenkins, Ian, *Archaeologists and Aesthetes in the Sculpture Galleries of the British Museum 1800-1939*. London: British Museum Press 1992, Chapter 7, pp. 140-152.

7. Newton, Charles, *Discoveries in the Levant*. London: British Museum 1865.
8. Stratford Canning, later Lord Ist Viscount de Redcliffe, had a long experience of Turkey. He first accompanied the Ambassador Robert Adair from 1808 to 1818 and then returned himself as Ambassador in 1826-27, and again from 1842 to 1858.
9. The Seven Wonders of the Ancient World are: the Great Pyramids of Giza, the Hanging Gardens of Babylon, the Statue of Zeus at Olympia, the Temple of Artemis at Ephesus, the Mausoleum at Halicarnassus, the Colossus of Rhodes, the Pharos at Alexandria. See Clayton, Peter and Price, Martin (eds) *The Seven Wonders of the Ancient World*. London: Routledge 1988 for an easily accessible account of them.
10. Pliny, *Natural History*, XXXVI, pp. 30-31. Loeb Classical Library, Harvard University Press 1938-62.
11. Luttrell, A. and Jeppesson, K., *The Maussolleion at Halikarnassus Vol II, The Written Sources*, Denmark, Moesgaard 1985, pp. 166-167.
12. Newton, Charles, *A History of Discoveries at Halicarnassus, Cnidus and Branchidae*. London, British Museum 1862, 2 vols. Vol 1, p. 325.
13. *See* Clayton pp. 105-106.
14. Dickson, pp. 30-32.
15. *See* Jenkins, Chapter 8, pp. 186-191 for a clear account of Newton's excavations.
16. ibid., pp. 24-25.
17. ibid., p. 33.
18. ibid., p. 36.
19. ibid., p. 57.
20. ibid., p. 59.
21. ibid., pp. 59-60.
22. ibid., pp. 353-367.
23. ibid., pp. 119-133.
24. Jeppesson, K., Hojlund, F., and Aaris-Sorensen, K., *The Maussolleion at Halikarnassus, Vol I, The Sacrificial Deposit*. Denmark: Aarhus 1986.
25. The Department of Antiquities of the British Museum was divided into the following sub-sections: Oriental Antiquities including the Egyptian and Assyrian, British and Medieval and Ethnographic Collections: Coins and Medals; Greek and Roman Antiquities.

26. Murdoch Smith, R. and Porcher, E.A., *A History of the Recent Discoveries at Cyrene made during an Expedition to the Cyrenaica in 1860-61*. London, British Museum 1864.
27. Scarce, Jennifer, 'Persian Art through the Eyes of Major-General Sir Robert Murdoch Smith KCMG', in *The Enterprising Scot* edited by Jenni Calder, Edinburgh HMSO 1986, pp. 131-138.

## CHAPTER ELEVEN
# Listening for 'The Sound of Running History'¹: Sir George Adam Smith (1856-1942)

## Iain D. Campbell

George Adam Smith's travels to Palestine must be studied in the light of his churchmanship. Son of a missionary of the Free Church of Scotland,² Smith grew up in a period of intellectual and ecclesiastical ferment. In writing of Smith's friend and colleague, Henry Drummond (whose biography Smith himself would eventually write), Professor Alec Cheyne does not exaggerate when he says that during the last 30 years of the 19th century:

> Scottish Presbyterianism's theological complexion changed more rapidly and profoundly than at perhaps any other time between the Reformation and our own day. The Free Church, in particular... quite suddenly surrendered its cherished early reputation as one of the most staunchly conservative and traditionalist bodies in Christendom.³

Borne along on a tide of imperialist expansion, theological revision and intellectual advance, Smith himself was not merely shaped by these changes – in many ways, as a biblical scholar within the critical tradition, he helped to forge them. His contribution to Scottish intellectual life was primarily in the areas of Old Testament exposition and the historical geography of Palestine. He produced biblical commentaries[4] and major works on the Holy Land,[5] as well as a massive range of articles on subjects as diverse as Scottish social conditions, the war and agricultural education. These writings reflect Smith's colourful career as Free Church minister of Queen's Cross Church, Aberdeen (1882-92), Professor of Old Testament at the Free Church College, Glasgow (1892-1910) and Principal of Aberdeen University (1910-1935).

This career must be interpreted within a changing ecclesiastical context. Smith's career began within a Presbyterian denomination committed to a 17th century Calvinistic confession of faith, with a high view of biblical inerrancy and inspiration, and a simple liturgy. Gradually, old wineskins came to be filled with new wine. The old Westminster Confession of Faith proved too restrictive for the late 19th century Free Church, which had come to question most of the fundamental doctrines of its heritage. The Bible, largely through the teaching of A.B. Davidson (1831-1902), Smith's Old Testament teacher at New College, Edinburgh, and W.R. Smith (1846-94), Davidson's famous pupil and Old Testament scholar, came to be regarded not as a uniquely inspired Scripture from God, but as the literary product of a developing religion. And the introduction of instrumental music and hymnody to the Scottish church marked the transformation not only of its substance but also of its form.

By the time Smith graduated from New College in 1880, ready to embark on his ministerial career, most of these changes were well underway, and received in him a willing advocate. Not the least influences on his thinking were his first-hand exposure to critical studies of the Old Testament at Tübingen (1876) and Leipzig (1878), studies which convinced him that the Old Testament is:

> not a set of dogmas, nor a philosophy, nor a vision; but a history, the record of a providence, the testimony of experience, the utterances called forth by historical occasions from a life conscious of the purpose for which God has called it and set it apart through the ages.[6]

The opportunity to travel, coupled with the desire to feel the 'sound of running history' which Smith felt pervaded the Old Testament, occasioned his four trips to Palestine. These, in turn, gave birth to some of his most lasting works on the historical geography of Palestine, works which are still cited by scholars in this field. We shall look in turn at his visits to Palestine and then assess his distinctive contribution.

*Egypt and Palestine*

Towards the conclusion of his New College studies, Smith received an invitation from a friend of his father, Dr Lansing, who was a missionary with the United Presbyterian Church of America, to accompany him on a trip to Cairo. The purpose of the trip was to help him with some mission work there, and it occupied him over the winter of 1879-80. A difficult sea journey in December 1879 took him en route to Lisbon and Algiers before arriving at Cairo at the end of the month. For Smith, every stop was both an adventure and an education, and his letters home are filled with lengthy descriptive prose which brought the images of these cities to life for his family. Of Algiers he wrote:

> What a strange sight... The squares are filled with palms, and among the crowd of strolling Frenchmen, who have brought over a sober edition of Paris life, move any number of tall broad-shouldered Arabs in their long white robes... How like ghosts they look in the shade of the palms. I suppose Algiers is what Cairo will turn out to be when I see it – a meeting place of East and West.[7]

One of the most significant – and little known – events of this trip occurred with the arrival of the ship, the SS *Canara*, at Port Said. Smith recounted the events to his mother:

> What was my surprise to see Robertson Smith step on board the 'Canara' and take possession of the bunk I had just vacated. He was on his way to Jedda, and thence to Aden to go to the interior of Arabia... I had time to have a chat with him...[8]

This chance meeting between George Adam Smith and William Robertson Smith may well have been the first time the two met for any length of time. The significance of the meeting, far away from ecclesiastical and theological controversies in Scotland, cannot be overlooked for, in the privacy of a vessel docked in Port Said, they had a unique opportunity to discuss events in which they were both involved.

William Robertson Smith was at this point engaged on a second visit to the East, from November 1879 to May 1880. According to his biographers, the effect of this trip was that he returned 'greatly invigorated in health, and bringing back with him a rich accumulation of observations and experiences that greatly influenced all his subsequent thinking and writing'.[9] A month in Cairo was spent learning Arabic and studying the religious experience of those among whom he travelled. The subsequent part of his visit was to Jeddah, before his return to Scotland in time for the crucial Assembly of 1880. Increasingly alienated from the Free Church leadership (as much a result of his acerbic personality as his views of the Old Testament), Robertson Smith was deposed in 1881 and subsequently taught at Cambridge.

As George Adam Smith recollected, looking back on the encounter with Robertson Smith in Egypt, 'He seemed buried under his pith-helmet... but I wish my hat enclosed what that helmet does'.[10] Clearly his respect for the Professor was great, and his meeting with him was one of the most formative experiences of his life, confirming in his mind the necessity of remaining in the Free Church in order to continue the work of critical scholarship. Interestingly, following Robertson Smith's suspension in 1880, it was George Adam Smith who was appointed to conduct his classes in Hebrew for the ensuing year.

George Adam Smith continued his visit to Egypt, remaining there for most of the spring. Part of the time was spent learning Arabic, under the tutorship of Socrates Spiro. This was Smith's first exposure to Oriental life and culture, and he found the experience fascinating. He wrote to his aunts:

> To live in Cairo is to live in a chop sea between two cross tides. The West and East meet here with no little confusion... People told me I should be disappointed with the East. I wasn't, only because I came here with not a single imagination about it. I had read nothing – or next to nothing, and Oriental life has always lain outside my conceptions.[11]

The sights of Egypt – broken down buildings left in a heap of rubble in the streets, large and ancient mosques, the howling of a crowd of people forming a funeral cortège, the sun setting behind the pyramids (with a transformation of landscape in glorious colour which, Smith says, 'I have only seen equalled on our islands off Ross-shire and in Skye'[12]) –

were all carefully noted by Smith, and encapsulated in the poetic prose style which characterises his letters. The Muslims he pitied; sensing in the silence of the mosques the Muslim apprehension of God, Smith concluded that 'they cannot pray except they look to Mecca, and they look to Mecca not because they find God there, but a sinful man like themselves'.[13]

Although in April 1880 Smith could write to his parents of leaving Cairo with very great regret at parting not only from the place but also from 'all the kind friends there, both natives and Americans',[14] the greater prize lay ahead. From Egypt Smith made his first visit to Palestine. It was a journey Smith was initially reluctant to make, for reasons which he does not express fully; a short postcard sent to his father in February 1880 says:

> Do you know I have a feeling against visiting Palestine?... Whatever I finally resolve upon, you will agree with me that it is foolish to cut short opportunities that I may never have again.[15]

Thankfully, Smith did not succumb to his feelings, and in Palestine he met with and spoke to many natives, in order to gain an understanding of Palestinian life and culture. The land, according to his wife's memoir,

> unfolded itself to him in all its many aspects. His eyes were opened to see the reasons and the meanings of words and events which, until then, he had only partly understood, and there is no doubt that the idea of writing a book upon the country was born in his mind from that first solitary journey.[16]

Whether or not the idea of a publication was conceived at this time, Smith was to state in the preface to the first edition of the *Historical Geography of the Holy Land*, that 'personal acquaintance with the land' was a necessary feature of the study of geographical and historical context. If he could confirm to his parents, following a visit to Bethlehem in late April 1880, 'how well the Bible story reproduces its general character',[17] it was because of painstaking, objective and extensive scientific study of the contours of Palestine.

The intrusion of foreign and non-Oriental nationals into Palestine filled him with loathing. The bazaars he described as 'horrible places, full of filthy flea-infested crowds. They are not Oriental. German watchmakers and carpenters, Polish shoe-makers, Spanish merchants... Russian tea-sellers...Americans in all sorts of dresses...'[18] He would also inform his readership that:

> In the Spring of 1880 I made a journey through Judea, Samaria, Esdraelon and Galilee, that was before the great changes which have been produced on many of the most sacred landscapes by European colonists, and by the rivalry in building between the Greek and Latin churches.[19]

The sense of disquiet pervades Smith's correspondence from the trip. Of the environs of Jerusalem and the supposed site of the crucifixion, where guides sought to point out the hole in which Christ's cross was placed, he writes: 'it is all saddening – the work done for mankind was so great, and the results are so very much the other way from that which the Maker intended.'[20]

There were, however, memorable moments, such as bathing in the Dead Sea ('my arms were aching from the intense efforts required to move through such dense water'[21]) and witnessing the sun rise over Jerusalem:

> I rose yesterday morning at half past four and rushed up the Mount of Olives in the soft warm dawn to see the sun rise over the mountains of Moab on the other side of the

Jordan valley... I enjoyed the sight of the city waking up exceedingly.[22]

George Adam Smith had himself woken up, alive now to the fascination of the Holy Land and the various aspects of its life and character.

Smith's second visit to Palestine was in 1891, the gift of a grateful congregation which had discovered that it was to retain the services of its minister, following his declining a call to Edinburgh. Now married to Lilian Buchanan, whom he had met in Switzerland, Smith and his wife travelled to Palestine in the company of Smith's former assistant, C.A. Scott, and P. Carnegie Simpson, future biographer of Principal Rainy. The expedition proper commenced in Egypt in March, with a visit to some of the famous sites there, following which the party explored Palestine. Thereafter they were zealous wanderers through biblical sites until they returned home in the autumn.

Lilian Smith's biography is painstaking in its detail on this expedition, emphasising the relative importance of this particular visit. Smith is portrayed as studiously surveying every minute detail of landscape:

> While the rest of us were interested in the general aspect and scenery, he was making notes of the details of each place, and at the same time his imagination, backed by his historical knowledge, called to life the events of the past, and envisaged the possible developments of the country in the future. He would see in his mind's eye terraces of vineyards upon a certain limestone ridge, or fighting taking place on a hill like Gezer. Never a day passed without his careful readings of the temperature and barometric pressure, and never an evening, however long and tiring the day, without his minute recording of everything done, seen, and heard – places, people, views and contours, vegetation, soil, animals, rocks, water, trees and flowers.[23]

At one point during their travels through the Judean desert, the party was ambushed by Arab tribesmen – about 80 of them, according to Lilian Smith's record – who made threatening gestures. 'One old villain tilted his long spear at George several times.' However, by remaining calm the party showed that if they were trespassing, which seems to have been the tribesmen's grievance, it was unintentional, and they were allowed to pass safely on their way to Jerusalem.

In Jerusalem the Smiths witnessed the religious celebrations of the Greek Orthodox Easter and explored the major biblical sites. These days in the city which Smith would describe as 'the bride of Kings and the mother of Prophets'[25] were, in his wife's reckoning, 'unforgettable'.[26] The party climbed Mount Hermon, then made their way to Damascus. There they stayed with friends who worked for the Edinburgh Medical Mission. It was part of Smith's intention during this trip to visit as many missionaries and mission stations as he could; the *Aberdeen Free Press* reported that he:

> called at nearly all the mission stations in Palestine, especially those of the Free Church; and he also visited the Edinburgh medical Mission at Damascus, and Dr Carslaw's at Sweir. What struck him most, he informed our representative, was the number of young missionaries, of from five to eight years' standing, who are at present in Palestine... In Mr Smith's opinion, no mission should be formed in a country like Palestine without having a medical section...[27]

Soon thereafter their two companions left them, and Smith and his wife continued to travel throughout the area east of the Jordan. Ancient Greek and Roman temples were explored, and 'George tired out all our men, whom he took with him in relays, to look for inscriptions'.²⁸ In Moab, through the influence of the Rev Henry Sykes, a missionary, they were guests at a Bedouin feast in the desert.²⁹

By the time Smith visited Palestine in 1891, he had already shown a professional interest in geography by joining the Scottish Geographical Society and becoming its secretary in 1885. In his study of the Royal Geographical Society, T.W. Freeman has argued that the period from about 1885-1895 was a decade not simply of growing interest in geography and geographical education; it was also the case that 'definition of the various aspects of geography was a natural preoccupation of the time along with the relation of geography to other subjects'.³⁰ In her study of Scotland as a seed-bed of academic geography, Elspeth Lochhead identifies as a distinctive Scottish contribution to geographical studies a 'holistic view of man and environment, which generally managed to avoid the pitfalls of extreme environmental determinism and related perspectives'.³¹ It is in this academic milieu that we find Smith developing his view of the geography of Palestine as a vital component of biblical studies, relating geography to historical theology and producing a *heilsgeographie* – a study of the geographical locus of salvation. It was not sufficient, in Smith's view, to ask of the biblical narrative 'what happened?' It was necessary to ask 'why did what happened just *then* also happen just *there*?'³²

The visit to the Holy Land gave Smith cause to lecture and write about his travels once he returned home. Over the winter of 1891/2, he delivered a series of lectures in Aberdeen on his visit, 'descriptive of the Holy Land, the missions there, and the Christian churches in the East'.³³ He was much in demand as a speaker at various other venues. In December 1891 he delivered a lecture in Huntly Free Church to the Women's Bible Study Assocation on 'The geography of Palestine in connection with the history of

Revelation', in which he described Palestine as 'a great bridge crossing between Egypt on the south and Assyria on the north, and enclosed on the one side by the Eastern desert and on the other by the Mediterranean Sea'.[34] Within the same week he addressed the Aberdeen Philosophical Society with a paper entitled 'Notes of a recent journey through the Hauran and Gilead, with a number of inscriptions discovered during its progress'.[35]

He also published, in 1891, a review of a guidebook which he says he took with him on the visit to Palestine. This was Baedeker's *Palestine*, first published in 1876. Smith's review points out some of the defects he found in the third edition of this volume, recently published. Smith argues that 'No adequate account is given of the Jordan valley, none at all of the east side of it. Pella, for instance, is altogether omitted; of Pella, on which so much has been written lately, the tourist is not even told that it exists... The archaeology is by no means up to date... and facts that stare the tourist in the face are not mentioned'.[36] Of significance in this review article, notwithstanding these criticisms, is the manner in which Smith describes what a 'manual of sacred geography' should be: a book to help determine:

> the distances and difficulties of biblical journeys, the lines of the ancient campaigns, and generally all the perspective of the Holy Land, as well as the latest results of biblical archaeology and geography.[37]

## Scholarly interest in Palestine

Smith's was an important voice in the late 19th century, reflecting a contemporary development of interest in European life and culture in general, and in biblical history in particular. John Bartlett sets the context for such interest thus:

The nineteenth century saw the dramatic expansion of archaeological and biblical study. This expansion owed much to political and economic factors such as the quest for a land route from the eastern Mediterranean to India, the imperial designs of Napoleon (whose surveyors mapped Palestine), the arrival of the steam ship and the steam locomotive, the development of photography and of a cheaper printing technology, and the growth of education for all. In an era when the Protestant churches set a high premium on biblical knowledge and Sunday Schools flourished, there was increasing interest in biblical geography, biblical peoples and their customs, and a ready market for the hundreds of books, especially illustrated books, published on Palestinian travel.[38]

This led to the development of historical geography as a literary genre. In surveying this development, Robin Butlin distinguishes between historical geography and geographical history. The first he defines as:

> the study of the geographies of past times, through the imaginative reconstruction of phenomena and processes central to our geographical understanding of the dynamism of human activities within a broadly conceived spatial context, such as change in the evaluation and uses of human and natural resources, in the form and functions of human settlements and built environments, in the advances in the amount and forms of geographical knowledge, and in the exercising of power and control over territories and people.[39]

Butlin defines geographical history as the 'study of the history and geography of changes in the territorial possessions and boundaries of states, empires and royal houses and administrations'.[40] He cites George Adam Smith as an example of the particularisation of the genre of historical geography in the area of biblical studies,[41] and describes the *Historical Geography of the Holy Land* as 'an important book whose intellectual context reflects wider currents of nineteenth- and early twentieth century thought'.[42]

This intellectual context included the attempt to establish an approach to history that was not conditioned by literary texts alone. In his study of the Annales school of historical geography, pioneered by Lucien Febvre and Marc Bloch in the early 20th century, Alan Baker has suggested that historical geography rose partly as a reaction against 'positivist history – which believed that the "hard facts" of historical reality were contained within documents'.[43] The result of this reaction was a holistic and synthetic study of history which included not only the evidence of literary texts, but evidence also from other disciplines. Baker notes that:

> the inseparability of geography and history, their combination in the form of regional histories, is a fundamental tenet of the Annales school and directly reflects its search for synthesis, its trust in total history. Such an holistic conception of the study of history was inevitably associated with advocacy of an interdisciplinary approach which soon extended beyond collusion with geography to cooperation with the full spectrum of academic disciplines, all of which were considered as potential handmaidens to history.[44]

Smith, though writing earlier than the Annales school, nonetheless applies similar principles of synthesis to the biblical history. The development of the historical geography of ancient cultures, not least the Palestine of scriptural history, grew out of a desire to relate the biblical source to external evidential sources, and thus produce a comprehensive view of the biblical world.[45]

Coupled with the intellectual rise of historical geography as an academic discipline was the practical issue that arose out of the new science of the 19th century. The Darwinian assault on the Bible led to a move to defend the literal truth of the Bible on the grounds of archaeology, which, some believed, would dig a grave for higher criticism.[46] As Machaffie observed:

> The fortresses of orthodoxy had been badly shaken in the past decades by the infidelity allied to political radicalism as well as the encroachment of historical and literary studies; here [in archaeology] was new hope which could be easily understood.[47]

Herbert Hahn pointed out long ago that this quest was bound to lead to disillusionment if such archaeological evidence was not forthcoming. He argues that in fact it did lead to disappointment in the early part of the 20th century, and suggests that at the back of it lay a wrong view of the function of archaeology itself. 'The real function of archaeology in relation to biblical studies,' he argues, 'is not confirmation but illumination. The goal is to understand the Bible, not to defend it.'[48] In studying the historical geography of the Holy Land, Smith and others were not so much concerned with defending the Bible against the new criticism, as with showing how criticism itself could defend the Bible against both an older rationalism and a conservatism which was inimical to the conclusions of modern science. For Smith, historical geography provided the illumination required for understanding the Bible.

For those who welcomed these new frontiers in biblical studies, the rediscovery of the Holy Land through a range of intellectual disciplines, would provide an objective, scientific, non-controversial and factual account of the Land of the Bible. An important development in this connection was the inauguration of the Palestine

Exploration Fund (PEF) in 1865. The Archbishop of York, in his address at the inaugural meeting of the PEF, stated the principles on which the work of the Society would be conducted:

1. That whatever was undertaken should be carried out on scientific principles.
2. That the Society should, as a body, abstain from controversy.
3. That it should not be started, nor should it be conducted, as a religious society.[49]

The intention to place on record only the facts discovered, and not an interpretation of them, proved extremely difficult. The Committee was resolved, however, 'to provide accurate information as to facts, leaving it to others to utilise those facts in whatever way they may consider desirable'.[50] The Fund's first major activities were based at Jerusalem. Although some of the conclusions of the Fund would remain open to debate, there is no doubt that the PEF did much to encourage a scientific study of biblical literature.

George Adam Smith was to work closely with the PEF, both as a member of its Council and a contributor to its *Quarterly Statement*. Indeed, as he looked back over the work of British excavations in Palestine, Smith could contrast it 'with the work of the excavations of other nations, and he knew that in thoroughness especially, the British work far excelled.'[51]

In her assessment of the archaeological work in Palestine in the 19th century, Naomi Shepherd says that the Palestine Exploration Fund 'had been established at the height of the passion for the biblical geography of Palestine which was already on the wane by the 1880s'[52]. She suggests that the mood in the 1880s was different, as scientific methodology had advanced and the wider world picture was emerging.

To concentrate on Palestine was to concentrate on a small corner occupied by a vassal kingdom against the backdrop of a wider and more important world.

However, this was precisely what fascinated Smith about the Holy Land, that in the light of the Old Testament, this small kingdom was superintended by 'the ideal of a special covenant between God and the Hebrew nation'.[53] To Smith the dominant religious ideology of the Scriptures was:

> That the God of this little tribe should be the Sovereign of earth and heaven!... Jerusalem asserted to be the centre of the whole earth, to which the Gentiles should bring their substance – Zion and Jordan exalted above all the hills and rivers of the world – Jews to be kings and priests to God, but the sons of the alien their plowmen and vinedressers![54]

For Smith the conjoining of Palestine's relative geographical isolation with its almost absurdly high profile in the Scriptures was the most attractive feature of his study. His was, consequently, not simply a geography of the Holy Land, but a *Historical Geography*, from which the Scriptural and theological aspects could not be omitted.

For Smith, study of Palestine was not a study in nationalism, but a study in Providence. It was not so much the study of the manners, customs and history of a people, as a study of the way in which that people had been covenanted in special relationship to God. This approach harnessed the testimony of the Old Testament writings and the witness of extra-biblical testimony to produce a theological naturalism[55] which sought to explore the relationship between the texts which spoke of God's election of Israel and the contexts in which this election manifested itself.

## Smith's Historical Geography of the Holy Land

Smith published his *Historical Geography of the Holy Land* (HGHL) in 1894.[56] By any standard, it was a remarkable piece of work, which included new maps and a comprehensive introduction to the geographical context of the biblical material. The preface to the first edition sets forth Smith's rationale for its publication. First, he says, 'the following chapters have been written after two visits to the Holy Land'.[57] His personal experience of travel, therefore, becomes, for the reader, a primary authentication of the book. Second, Smith draws attention to European involvement in Palestine: 'the real exploration of Palestine', he says, 'has been achieved during the last 20 years',[58] not only in the many activities of the PEF, but also in the labours of other societies and individual explorers.

Most importantly, Smith argues that 'an equally strong reason for the appearance at this time of a Historical Geography of Palestine is the recent progress of Biblical Criticism'.[59] He claims that his is the first historical geography to pay heed to the two-fold duty of serious students of Scripture: the duty, first, of 'regulating the literary criticism of the Bible by the archaeology of Syria', and the converse duty of showing the 'helpfulness of recent criticism' in writing the geography of the Holy Land.[60]

Smith's work is divided into three books. Book 1 deals generally with the land, in its relation to the history of the world, its form, climate and scenery, and also the questions of faith which relate to aspects of the geography of Palestine. Book 2 deals more particularly with Western Palestine, and Book 3 with Eastern Palestine. Five appendices deal with geographical passages and phrases in the Old Testament, the theory of Israel's invasion of Western Palestine, the wars against Sihon and Og, the bibliography of Eastern Palestine, and the subject of roads and vehicles in Syria.

Smith operates from the principle that as a theological concept, geography may have a symbolic or spiritual significance, so that 'geographical categories are being employed, but their literal geographical reference has completely disappeared. The geographical has become the servant of the spiritual'.[61] In this way Smith can describe the 'Old Testament pictures of landscape' as 'testimonies of the truth of the narratives in which they occur',[62] and therefore as aids to faith. In such a context, Smith says, 'the geography of Syria exhausts the influence of the material and the seen, and indicates the presence on the land of the unseen and the spiritual'.[63] Or, to put it at its most stark, for the Christian, in Smith's view, 'his Bible is [Palestine's] geography from Beersheba to Antioch'.[64]

The sermonic style is evident throughout, as Smith gives a descriptive narrative followed by a statement of its theological significance. Geographer Robin Butlin assesses the work as:

> a pioneer book written by a powerful communicator and preacher, whose primary concern was to evoke and use an array of modern scientific and critical techniques for an essentially evangelical purpose, but whose intellectual milieu... reflects wider currents of thought and ideas.[65]

Butlin explores the influences on Smith's thinking – the Bible as a primary source, the changing face of Palestine, and the development of the study of geographical context, as well as the intellectual connections which enabled Smith to produce his pioneer volume. He sets Smith's work in the context of 'a watershed or divide in work on the historical geography of the Holy Land',[66] standing midway between works which relied for the most part on secondary sources, and later works which would incorporate a wide corpus of new scientific discoveries. At least one critic recognised this when the HGHL appeared, stating that, 'it is, of course far from completing its task; it is really only the first opening up of what will hereafter prove a fruitful field of study'.[67]

For Smith, writing as a churchman, his chapter on 'The Land and Questions of Faith' is important. He states as a guiding principle the following: 'That a story accurately reflects geography does not necessarily mean that it is a real transcript of history... let us at once admit that, while we may have other reasons for the historical truth of the patriarchal narratives, we cannot prove this on the ground that their itineraries and place-names are correct.'[68] On the other hand, he maintains that geographical descriptions authenticate biblical passages 'as testimonies of the truth of the narratives in which they occur'.[69]

This is all very well; but Smith's acceptance of the conclusion of higher criticism on the question of authorship of biblical texts rather begs the question. He admits that issues related to authorship must be decided on grounds other than geographical ones; but his suggestion that the descriptions of Jerusalem in proto-Isaiah, which he contrasts with those of deutero-Isaiah (without offering evidence) assumes rather naively that 'the evidence of geography mainly comes in support of a decision already settled by other proofs'.[70]

For Smith, the geography of the Holy Land lays the foundation for a developmental view of Israel's religion. Although geographical study cannot fully answer all the questions that arise in connection with literary criticism of the Old Testament, he assures his readers that 'when we rise to the higher matters of the religion of Israel, to the story of its origin and development, to the appearance of monotheism, and to the question of the supernatural...the testimony of the historical geography of the Holy Land is high and clear'.[71] Again, however, Smith assumes that Israelite religion was originally polytheistic. He replaces the view of the French critic Renan concerning Israel's primitive monotheism, with the critical view that 'in the Semitic religion, as in the Semitic world, monotheism had a great opportunity'.[72] Later he argues that the fertility of the land invited polytheism, and made monotheism highly unlikely.[73]

The geography, therefore, has become a foil for the revelation; and the emergent monotheism has developed not because of, but in spite of, the geographical position of Israel. Neither assumption – either that the fertility of the land necessarily inclined to polytheism, or that the monotheistic religion of Israel was a late development – is evident on *a priori* grounds. Smith's argument in the HGHL is that the physical nature of Israel would have led to the Semitic nation being polytheistic were it not for the revelation which they received from God.[74] But this appears to be a highly subjective and circular argument.

*Further travels to Palestine*

Smith visited Palestine for a third time in April 1901, by which time he was Professor of Old Testament language and literature at the Free Church College, Glasgow.[75] He journeyed in the company of four ministerial colleagues, five students and a Glasgow businessman.[76] Lilian notes that travel through Palestine was much improved since the last visit, with railways and new roads opening up the country; one result of this was a concentration of travel east of the Jordan. Smith was disappointed, however, that the progress in building on ancient sites was hampering the archaeological work necessary for the study of ancient cultures. In the publication of his notes in the Palestine Exploration Fund *Quarterly Statement* for 1901 Smith urges that 'every year means irrecoverable loss',[77] a point which he uses to urge increased subscription to the Fund.

Smith's notes indicate a strong interest in military campaigns in Palestine. One of the less well known aspects of this journey was a discovery at Tell Esh Shihab, in Hauran, some 30 kilometres east of the southern tip of the

Sea of Galilee. Smith concluded from the low elevation of the area, and the water courses in and around Tell Esh-Shihab that it 'must always have been a site of great importance';[78] and his discovery of a monument of Sety I of Egypt there corroborated his view. Smith concluded that the stone was 'of no little importance in connection with the conquests of the Pharaohs on the east of Jordan'.[79] The prevailing view at the time was that Egyptian conquests of Palestine had not extended much into western Syria; but Smith challenged that on the basis of his find, concluding that Sety had in fact crossed the Jordan and extended his conquest into eastern Palestine.

The Palestine Exploration Fund *Quarterly* also published Smith's extensive notes of his fourth and final visit to Palestine, in 1904. This visit was the result of Smith's having contracted a serious illness during a visit to America in 1903, which left him unable to fulfill his teaching duties during the 1903-4 session. He was advised to travel to recuperate, and a financial gift from some friends in Glasgow made it possible for him to visit India, his first visit to the country of his birth since 1858. Smith's wife accompanied him to India, but the exploration of Palestine was in the company of her brother, Dr G.S. Buchanan. The journey focussed on Moab, east of the Dead Sea, after which Smith and Buchanan returned to Jerusalem. There Smith spent much time examining the ancient city of David, particularly, according to his wife's memoir, 'investigating water-courses and walls and subterranean passages and inscriptions, in all of which he received much interested assistance from authorities living in the city. He had in view his book upon Jerusalem, and this was a most valuable and helpful visit'.[80]

## Jerusalem

The main lacuna in the HGHL was that there was no treatment of Jerusalem.[81] A series of articles by Smith appeared in *The Expositor* between 1903 and 1906, which were later reprinted with additional material in two volumes entitled *Jerusalem: the Topography, Economics and History from the Earliest Times to AD 70*. The first volume appeared in 1907, the second in 1908. They were in a sense even more of a pioneering work than the HGHL, since 'hitherto there had been no specific study of Jerusalem'.[82]

While the principles governing *Jerusalem* were the same as those governing the HGHL, the focus of interest was much narrower. Volume I divides into two books, Book 1 dealing with topographical subjects (sites and names), and Book 2 with the economic and political background to Jerusalem. Volume II contains Book 3, which is a historical study of Jerusalem (and which his wife, Lilian, assessed as 'one of the best pieces of work George ever did'[83]).

The introductory essay in Volume I, 'The Essential City' is a summary of the place and significance of Jerusalem. A superb piece of writing, it introduces the reader to the main points of importance of the city who 'knew herself chosen of God, a singular city in the world, with a mission to mankind'.[84] Though inferior to other places in learning and philosophy, Jerusalem became 'the home of the Faith, the goal of most distant pilgrimages, and the original of the heavenly City, which would one day descend from God among men'.[85] In these introductory remarks, Smith recalls some of his travels around Jerusalem, and shares with his readers his memory of the scenes which he witnessed. These serve as a fitting introduction to his description of the city; and the geography is once again invested with symbolic and spiritual significance.

Book 2, on the economics and politics of Jerusalem, examines the growth of the capital at the expense of the provinces. Jerusalem evolved into the only legitimate place of sacrifice-worship and the place of pilgrimage for many of the faithful. Given its physical elevation, Smith asks, 'how were her finances regulated, and whence did she draw provision both for so numerous a non-productive population and for the temporary but immense additions to it caused by the Temple festivals?'[86] There follows a detailed examination of Jerusalem's natural resources, and an examination of the commercial imports, temple revenues, royal trade, crafts and industries which, in a city most unfitted, in Smith's words, 'to be the home of industries',[87] nonetheless acknowledged that all its resources it owed to the benevolence of God.[88] A discussion of the government and policing of the city and a note on the term 'the multitude', the common people, who laid the foundation for the Christian church of the New Testament concludes this section.

While touching briefly on ancient letters and documents relating to Jerusalem, for Smith the history proper begins with David's conquest, and with the religious impulse which inspired David to bring the Ark of the Covenant to the city. 'The national unity,' Smith says, 'had never been maintained, or when lost had never been recovered, except by loyalty to the nation's one God and Lord. His Ark implied Himself. It was His Presence which sealed the newformed union, and consecrated the capital.'[89] Smith credits David with making her the religious centre which she became, even if it was the prophetic word and the deuteronomic legislation which were the chief factors in her development: 'The Man, whose individual will and policy seem essential to the career of every great city, Jerusalem found in David... The drama of Jerusalem is never more vivid than while David is its hero.'[90]

Smith discusses the place of Jerusalem in the major prophets, the statesmen who safeguarded the role of Jerusalem as the city chosen by God. Elsewhere Smith describes Jerusalem as the focus of Isaiah's great interest: 'She is his immediate and ultimate regard.'[91] In typical prose style, Smith says that 'the fires which David and Solomon kindled in Jerusalem... leap into high, bright flame at the powerful breath of Isaiah'.[92] While other prophets spoke the divine word to the city, it was Isaiah's contribution that secured, amid the political upheavals through which she passed, a 'mind to read her history and proclaim her destiny'.[93]

In a chapter on 'The Ideal City and the Real', Smith examines the biblical literature which represents Jerusalem as the ideal of religious life. The exilic poetry and narrative mourned the loss of the city and anticipated a return. However, the later prophets, he argues, turned their attention from king to priest as the pivotal figure in Jerusalem's life, so that the return after the exile was not of a kingdom, but of a colony.[94] In the Second Temple period, Smith finds in Israel on the one hand a lofty idealism that God had somehow returned to his place in Jerusalem, but on the other an increased superstition – 'the contradiction of the idea that He dwelt only there';[95] from the sacredness of the Temple there developed 'the dogma of its inviolableness',[96] and the consequent superstitious confidence that God would never leave it. Not until the coming of Jesus was this superstition dealt with definitively, for 'the Messiah promised to the Temple supplanted the Temple'.[97] The crucifixion of Jesus represented the gross ignorance of Jerusalem of her religious position and heritage – it was the final confrontation of the City and the Man, Smith says;[98] nevertheless what was 'a sunset to herself' was 'the dawn of a new day to the world beyond'.[99]

Unlike the HGHL, *Jerusalem* contained photographs as well as maps. Many of these were taken by Smith himself, and made the book more appealing. These photographs were an important contribution to the academic study of the Holy Land. In her study of the role of photographs in disseminating geographical information, Joan Schwartz (1996) has argued that photographs had a wide appeal:

> Through travel photographs, in concert with other forms of representation, Victorian viewers who had never travelled came to share impressions of place. Not only as a

pool of visual facts, but also as symbols of imperial expansion, colonial development, commercial enterprise, military might, and scientific knowledge, these mutually-held visual images contributed to national identity, stimulated patriotic effort and reinforced one's sense of place in the world.[100]

Schwartz's view that 'the camera, like the pen and the brush, when wielded by Western travellers, depicted the world in Western terms',[101] may not be entirely discordant with Smith's own view of Jerusalem. Recognising that 'East and West hotly contended for her',[102] Smith argues that the physical location of Jerusalem suggests that 'Providence had bound over the city to eastern interests and eastern sympathies'.[103] His work may be seen as a liberating of the essential city from its eastern aspect, an attempt, as he puts it, to 'bring the spell with him out of the history'.[104] It is arguable that, for his Victorian audience, Smith was staking a Western claim in his writings.

In Butlin's view, Smith's analysis of Jerusalem's history is 'a remarkable combination of scholarly erudition, intuition, historical imagination and depth of feeling'.[105] Indeed, Smith's work on Jerusalem is cited in modern works on the same theme.[106] James Denney, the distinguished theologian and Smith's colleague at the Free Church College in Glasgow, to whom Smith gave a copy of the Jerusalem volumes, also drew attention to the admirable mix of qualities which fitted Smith eminently for the work of writing on Jerusalem:

> He has the eye of the geographer or the military engineer for the physical features of a situation, the vivid imagination by which the historian recognises great events, the enthusiasm and penetration of the prophet who discerns and interprets the spiritual crises in a nation's life. Without this combination of gifts and interests, which keeps his own

mind alive at every point, so thorough a book must have become intolerably heavy; as it is, it can be read from beginning to end with no less delight than gain.[107]

One of the disappointing features of the *Jerusalem* volumes was that Smith was unable to extend the history to the limits of the title. He did not, in fact, bring the history up to AD 70. This was pointed out in one review which said that 'the natural conclusion would have been the siege and destruction of Jerusalem by Titus; and it was planned to end with this; but the concluding chapter together with an appendix have been crowded out by the size to which the second volume has grown'.[108] The failure to reach the designed omega-point shows the massive amount of work which brought the study up until the early narrative of the Book of Acts. The volume of writing, presumably, also prevented fulfilment of W. Robertson Nicoll's wish of seeing *Jerusalem* and the HGHL published together in two volumes.[109] Desirable as this may have been, it is arguable that in their form and substance they are different; the *Jerusalem* volumes for example, are more Bible-oriented than the HGHL. It is probably better that they stand on their respective merits.

*Cartography*

The maps in the HGHL and *Jerusalem* were the work of John George Bartholomew (1860-1920), whose association with Smith was a major contribution to 20th century cartography and the development of the Bartholomew map-making business. In thanking Bartholomew for his help with the maps, Smith claims that they were the first orographical maps of Palestine – that is, maps which showed the mountainous elevations by using different colours for different heights,[110] instead of simply using varieties of black and white shading. Prefacing the

*Jerusalem* volumes, Smith thanked Bartholomew 'for all the trouble he has taken in their preparation, as well as for the clearness and impressiveness with which they have been achieved'.[111]

Towards the end of the 19th century, Bartholomew's was, in Elspeth Lochhead's words, 'a real focus of activity, having strong links with the scientific community centred in Edinburgh'.[112] Links and connections were important in Smith's works, as Butlin argues:

> The important connections are those which concern, for example, developments in the nature of historical geography and the scientific and artistic exploration of Palestine, but wider connections existed with other geographers and geographical and scientific societies, and [Smith's] work may also be viewed in the context of major debates on the critical appraisal of the Bible as historical evidence and on the compatibility of religion with Darwin's theory of evolution.[113]

The connection with Bartholomew was itself a door into an important scientific milieu which gave Smith an interest in, and a relevance to, a wide variety of disciplines.

Eventually Hodder and Stoughton published, in 1916, Smith's *Historical Atlas of the Holy Land*, planned by himself and Bartholomew as far back as 1894.[114] Butlin calls into question the accuracy of one or two details in the atlas, suggesting that the length of time which elapsed between initial planning and final publication did not allow for modernisation of Smith's first maps.[115] Nonetheless, it was to be a major landmark in the Bartholomew tradition, praised by the Bartholomew historian Gardiner as 'an astonishing compendium of religious history',[116] which:

> emerged unique, owing nothing to what had gone before, an excellent example of the selective thoroughness and

artistic refinement which John George preached, and the accurate up-to-date treatment which had come to be associated with his name.[117]

## Smith's contribution to Historical Geography

S.A. Cook is probably going too far when he says that Smith's work in the field of historical geography was 'revolutionising'.[118] In a more sober assessment, A.B. Bruce recognised that there were other studies of aspects of Palestinian geography which would answer more detailed questions.[119] Nonetheless he was ready to admit that there was something distinctive in Smith's contribution:

> Not antiquarian investigation into the claims of particular spots to be sites of historic towns, not a running commentary on biblical texts, not photographic pictures of what can be seen from selected viewpoints... but a comprehensive idea of Palestine as a whole, with careful description of its separate parts in their organic relation to the whole, and in connection with the historic drama enacted on the soil.[120]

As a geographer, Smith was able to give a comprehensive and holistic view of matters relating to the physical aspects of biblical history. His strength lies in his ability to bring together a wide range of academic interests and weave them into a whole. The passionate interest in geography, which grew out of his travels to the Holy Land, provided him not only with a frame of reference for biblical studies, and a background to the biblical literature, but also became a handmaid to his linguistic and theological interest in the Old Testament.

Modern writers are still ready to acknowledge the important role of Smith in the development of biblical geography. O. Palmer Robertson (1996), for example, speaks of Smith's HGHL as popularising a technical treatment of the subject,[121] while Max Miller, in a recent edition of *Biblical Archaeologist* could write that the pioneers of the discipline:

> made significant headway in clarifying the historical geography of Palestine. Edward Robinson's *Biblical Researches in Palestine and the Adjacent Regions* and George Adam Smith's *The Historical Geography of the Holy Land* are not just classics; they contain most of what we know today about Palestinian toponymy – including ancient place-names and the approximate or specific locations of biblical cities and villages. This sort of information is basic for any attempt to reconstruct the history of biblical times.[122]

This view holds that Smith's work is foundational to subsequent treatments of the historical geography of Palestine, with writers such as Y. Ben-Arieh talking of the HGHL as a 'classic' and demonstrating its 'popularity and timelessness' from the number of subsequent editions.[123]

However, a recent school of thought has questioned George Adam Smith's approach. Keith Whitelam, for example, while acknowledging the important work done by Smith, questions the imperialist and colonialist treatment of the Holy Land by his – and similar – studies. For Whitelam, Palestine, in Smith's treatment, 'has no intrinsic meaning of its own, but provides the background and atmosphere for understanding the religious developments which are the foundation of Western civilization'.[124] The HGHL Whitelam describes as 'a classic Orientalist expression of Europe's Other',[125] in which there is no real or authentic 'history', since the indigenous Palestinian history is silenced in the interests of 'Israelite' history, and the history of Western monotheism.

This thesis is part of a larger discourse among scholars as to the extent to which ancient Israel is an authentic historical concept. For Philip R. Davies (1992), for example, a major contributor to this discussion, 'ancient Israel' is a scholarly construct, literary in form, and sometimes given a vague geographical and historical setting in the Palestinian world (of whose history we know almost nothing). Davies accuses such scholarship of 'a retrojective imperialism, which displaces an otherwise unknown and uncared-for population in the interests of an ideological construct'.[126] Both Davies and Whitelam take issue with Smith's reading of the biblical texts and citing these as primary witnesses to the biblical history.

In spite of this position, it is still a valid argument that the biblical texts themselves ought to be integrated into a reconstruction of the history of Palestine. The observation of V. Philips Long is still pertinent:

> The social sciences can be useful... in pursuing questions that the text does not address, or does not address directly. But are we well advised to seek to escape the *constraints* of the text in matters that it *does* address?[127]

While it is true that Smith's approach may be viewed as part of an imperialist discourse, the approach of Whitelam and Davies seems to call into question the legitimacy of Smith's attempt to reconstruct a historical geography of the Holy Land. It is by no means obvious that Smith's studies and contributions to *historical* geography were informed merely by his Western imperialism. Smith's visits to Palestine were an attempt to feel the beat of the ancient civilisations. And to the extent that he wrote about what he saw, and also marshalled the evidence of the contemporary biblical writings as he understood them through the tools of modern criticism, his was not an attempt to silence Palestinian history in the interests of a merely academic construct of ancient Israel.

And at last, any study of Smith's travels brings us back to

where we began – to his churchmanship. If the Bible was a primary textbook for Smith in his exploratory travels, then the land itself became for him a context in which to interpret it. What Smith says of the Old Testament prophet Hosea is no less true for himself, that 'his sacraments are the open air, the mountain breeze, the dew, the vine, the lilies, the pines; and what God asks of men are not rites nor sacrifices, but life and health, fragrance and fruitfulness, beneath the shadow and the Dew of His Presence.'[128] The legitimacy of the approach may be open to question, but there is no doubt that Smith went to Palestine to see, and consequently, to believe. His travels in Palestine are paralleled by his search for a world view, which he found at the meeting-place of ancient civilizations and amid the scenes of an ancient faith.

## REFERENCES

1. The phrase is taken from Smith's preface to the first edition of the *Historical Geography of the Holy Land* (London, 1894).
2. Smith's father was George Smith, who taught in Doveton College, Calcutta from 1853-59, when he became editor of the *Friend of India* newspaper. He subsequently served as Secretary to the Foreign Mission Board of the Free Church of Scotland. It was in Calcutta that George Adam Smith was born, in 1856.
3. A.C. Cheyne, 'The Religious World of Henry Drummond (1851-97)' in *Studies in Scottish Church History*, Edinburgh, 1999, p. 185.
4. On Isaiah (2 vols, 1888 and 1892), the minor prophets (2 volumes, 1896 and 1898), the Book of Deuteronomy (1918) and Jeremiah (1923).

5. *Historical Geography of the Holy Land*, 1894, *Jerusalem*, 2 volumes, 1907 and 1908 and the *Atlas of the Historical Geography of the Holy Land*, 1915.
6. GAS, *Our Common Conscience*, London, 1918, p. 325.
7. GAS to parents, Algiers December 1879, NLS Acc 9446, No 16 [This refers to the major collection of Smith's papers, located in the National Library of Scotland, Accession Number 9446. These were thoroughly catalogued by Smith's daughter, Janet Adam Smith].
8. GAS to mother, Cairo, 21 December 1879, NLS Acc 9446, No 16.
9. J.S. Black and G.W. Chrystal, *The Life of William Robertson Smith*, London, 1912, p. 333.
10. GAS to mother, Cairo, 21 December 1879, NLS Acc 9446, No 16.
11. GAS to aunts from Cairo, 30 December 1879, NLS Acc 9446, No 16.
12. ibid.
13. ibid.
14. GAS to mother from Middle East, 11 April 1880, New College Library Special Collections, SMI 1.4.1.
15. GAS's postcard to his father from Thebes, 16 February 1880, NLS, Acc 9446, No 16.
16. Lilian Adam Smith, *George Adam Smith: A Personal Review and Family Chronicle*, London, 1942 (hereinafter LAS), 18
17. GAS to parents, 21 April 1880, New College Library Special Collections, SMI 1.4.1.
18. ibid.
19. G.A. Smith, *Historical Geography of the Holy Land*, preface to the first edition (1894), reprinted in the 22nd edition, p.xi.
20. GAS to parents, New College Library Special Collections, SMI 1.4.1.
21. GAS to parents, Monday 19 April, ibid.
22. ibid.
23. LAS, p. 49. The detail of Lilian Smith's account suggests that she too was busy with her pen; and there survives a record from her hand and privately printed of the visit to Palestine (entitled *East of the Jordan*, this small pamphlet survives in a journal of press cuttings about Smith in NLS Acc 9446 No 23).
24. ibid., p. 51.

25. GAS, *Jerusalem: The Topography, Economics and History from the earliest times to* AD *70*, Vol. I, London, 1907, p. 4.
26. LAS, p. 52.
27. Press cutting from the *Aberdeen Free Press* n.d. in journal of press cuttings about Smith in NLS Acc 9446 No. 23.
28. LAS, p. 56.
29. In the preface to the first edition of the *Historical Geography of the Holy Land* (HGHL), Smith expresses regret that his second visit to Palestine was not as complete as he might have wished: 'Unfortunately – in consequence of taking Druze servants, we were told – we were turned back by the authorities from Bosra and the Jebel Druz, so that I cannot write from personal acquaintance with those interesting localities, but we spent the more time in the villages of Hauran and at Gadara, Gerasa and Pella, where we were able to add to the number of discovered inscriptions' (p. xi).
30. T.W. Freeman, 'The Royal Geographical Society and the Development of Geography' in E.H. Brown (ed.) *Geography Yesterday and Tomorrow*, Oxford, 1980, p. 18.
31. E.N. Lochhead, 'Scotland as the Cradle of Modern Academic Geography in Britain', *Scottish Geographical Magazine*, 97 (1981), p. 108.
32. Freeman, 'The Royal Geographical Society', p. 26.
33. Press cutting from the *Aberdeen Free Press* [?] n.d. in journal of presscuttings about Smith in NLS Acc 9446 No. 23.
34. *The Huntly Express*, 12 December 1891, in journal of press cuttings about Smith in NLS Acc 9446 No. 23. The metaphor of Syria as a bridge is explored in Smith, HGHL, p. 6.
35. Page 15 of journal of press cuttings about Smith in NLS Acc 9446 No. 23. The notice is of a meeting on 15 December 1891. The synopsis of the paper notes the inscriptions discovered by Smith: 'Emperor Otho (69 AD), King Agrippa II (81 AD), Roman legion etc., Sarcophagi, pre-Christian and Christian inscriptions'.
36. GAS, 'The New Edition of Baedeker's *Palestine*, *The Expositor* (1891, Vol. IV, Fourth Series)', pp. 467-8.
37. ibid., p. 467.

38. J.R. Bartlett, 'What has archaeology to do with the Bible – or vice versa?' in John R. Bartlett (ed.), *Archaeology and Biblical Interpretation*, London, 1997, pp. 3-4.
39. R.A. Butlin, *Historical Geography: Through the Gates of Space and Time*, London: 1993, p. 1. Butlin's opening chapter is an interesting historical survey of the field of historical geography; he claims that his is the first attempt to do so.
40. ibid., p. 12, although Butlin concedes that this field of study was also described in some places as historical geography.
41. Butlin says: 'In Europe the concept and practice of 'historical geography' in the seventeenth and early eighteenth centuries, was closely associated with scriptural or biblical geographies of the Old and New Testaments. This then, as will be shown, continued to figure in historical geography throughout the nineteenth and the early twentieth centuries' (ibid., p. 2).
42. Butlin, *Historical Geography*, p. 7.
43. Alan R.H. Baker, 'Reflections on the relations of historical geography and the Annales school of history' in A.R.H. Baker and D. Gregory (eds.), *Explorations in Historical Geography: Interpretative Essays*, Cambridge University Press, 1984, p. 4.
44. ibid., p. 7.
45. cf. Walter C. Kaiser, 'The Current State of Old Testament Historiography' for a summary of some of the fallacies that have arisen out of this approach, such as that 'history cannot include anything that does not have external documentation.' (W.C. Kaiser, *A History of Israel from the Bronze Age to the Jewish Wars*, New York, 1998, p. 5). Kaiser is aware of the need for a holistic approach, but wishes to claim the biblical texts themselves as primary witnesses for Old Testament historiography.
46. See especially B.J.Z. Machaffie, *The People and the Book: A Study of the Popularization of Biblical Criticism in Britain, 1860-1914'*, PhD University of Edinburgh, 1977, especially chapter 5 'The Stones cry out: Archaeology and the Higher Criticism' pp. 501ff.
47. Machaffie, *The People and the Book*, p. 515.
48. Herbert F. Hahn, *The Old Testament in Modern Research*, London: SCM Press, 1956, p. 187
49. C.M. Watson, *Palestine Exploration Fund: Fifty Years' Work in the Holy Land; a Record and Summary 1865-1915*, London, 1915, p. 18. Although

some looked on archaeology as championing the old orthodoxy, it was not primarily for that reason that the PEF was established. The primary interest of the emergent society was scientific, though undoubtedly in the interests of objective biblical scholarship.
50. ibid., p. 19.
51. Report of address by GAS to the Glasgow branch of the Egyptian Research Students' Association, in *Glasgow Evening Citizen*, 26 October 1909.
52. N. Shepherd, *The Zealous Intruders: The Western Rediscovery of Palestine*, London, 1987, p. 226.
53. GAS 'The Ethiopian and the Old Testament: A Sermon on Acts 8:26-40' in *The Expository Times*, Vol. I, no. 10 (July 1890), p. 234.
54. ibid.
55. I have coined this phrase to describe the kind of enquiry Smith was conducting into Palestinian history. The phrase should not be confused with 'natural theology', which is a specific branch of theological study.
56. In his tribute to Smith's work, Yehoshua Ben-Arieh dates the HGHL wrongly, stating that it was written after the first visit in 1880 (he also gives a wrong date of 1899 for the second visit). Ben-Arieh includes the HGHL even although it falls outside the time parameters of his study, since he regards Smith's name as 'inseparably linked with the concept of the historical geography of the Holy Land'. (*The Rediscovery of the Holy Land in the Nineteenth Century*, Jerusalem, 1983, p. 225).
57. HGHL, p. x.
58. ibid., p. xi.
59. ibid., p. xvi.
60. ibid.
61. Grogan, G.W., 'Heilsgeographie: Geography as a Theological Concept', *The Scottish Bulletin of Evangelical Theology*, Vol. 6, number 2 (Autumn 1988), p. 89.
62. HGHL, p109.
63. ibid., p76.
64. GAS, *Syria and the Holy Land* (London, 1918) p. 6.
65. R. Butlin, 'George Adam Smith and the historical geography of the Holy Land: contents, contexts and connections', *Journal of Historical Geography*, 14.4 (1988), pp. 387-8.
66. ibid., p. 395.

67. W.M. Ramsay, 'Professor G.A. Smith's *Historical Geography of the Holy Land*', *The Expositor*, 5th series, Vol. I (1895), p. 57.
68. HGHL, p. 108.
69. ibid., p. 109.
70. ibid.
71. ibid., p. 111.
72. HGHL, p. 30. Smith also took issue with Renan's view at the Summer School of Theology in Oxford in 1894, when he lectured on 'The Preparation for Prophecy'. According to A.S. Peake's reporting of Smith's lecture in the *Methodist Times* for 9 August 1894, Smith demonstrated 'not simply that the tendency to monotheism did not exist, but that even those causes which favoured such development were unable to produce it among the Semites. The monotheism of Israel was traced to the direct revealing activity of God'.
73. ibid., p. 90.
74. ibid., p. 113.
75. And facing a trial for heresy over the publication of his lectures *Modern Criticism and the Preaching of the Old Testament*. In the event, the General Assembly of 1902 voted not to proceed to trial.
76. LAS p. 81: 'he had as his companions four young ministers, John Kelman, W.L. Robertson, Edward Roxburgh and Nicol MacNicol; five students, John Mackay, Donald Cameron, Candlish, Paterson and Hartzell; and a Glasgow business friend, Mr Arthur Hart.'
77. PEF *Quarterly Statement*, 1901, p. 341.
78. ibid., p. 344.
79. ibid., p. 349.
80. LAS, p. 90.
81. In the preface to *Jerusalem*, Vol. I, p. vii, Smith states: 'In the *Historical Geography of the Holy Land* it was not possible, for reasons of space, to include a topography of Jerusalem, an appreciation of her material resources, or a full study of the historical significance of her site and surroundings.'
82. Cook, 'Sir George Adam Smith', p334.
83. LAS, p106. Some 35 years later, Smith's friend and colleague, David S. Cairns, described the Jerusalem volumes as 'indispensable to the serious

professional student of the Bible' ('Sir George Adam Smith', *Religion in Life*, Vol. XI Number 4, (Autumn 1942), p. 537.
84. GAS, *Jerusalem*, Vol I, p. 4.
85. ibid., p. 7.
86. ibid., p. 276.
87. ibid., p. 372.
88. cf ibid., p. 374: 'the good artificer is not despised in the Old Testament; on the contrary, his gifts are regarded equally with those of the husbandman as from God'.
89. *Jerusalem*, Vol. II, p. 38.
90. ibid., p. 47.
91. *Isaiah*, Vol. I, p. 22.
92. *Jerusalem*, Vol. II, p. 147.
93. ibid.
94. See also Smith's *Book of the Twelve Prophets*, Vol. II, p. 189: 'Israel is no longer a kingdom, but a colony. The state is not independent: there is virtually no state. The community is poor and feeble... We miss the civic atmosphere, the great spaces of public life, the large ethical issues. Instead we have tearful questions, raised by a grudging soil and bad seasons, with all the petty selfishness of hunger-bitten peasants.'
95. *Jerusalem*, vol. II, p. 311.
96. ibid., p. 312.
97. ibid., p. 521.
98. ibid., p. 578.
99. ibid., p. 579.
100. Joan M. Schwartz, '*The Geography Lesson*: photographs and the construction of imaginative geographies', *Journal of Historical Geography*, 22.1 (1996), p. 31.
101. ibid.
102. *Jerusalem*, Vol. I, p. 7.
103. ibid., 11.
104. ibid., p. 22.
105. Butlin, 'George Adam Smith', p. 389.
106. Such as Auld, G., and Steiner, M. *Jerusalem I: From the Bronze Age to the Macabees (Cities of the Biblical World)*, (Cambridge, 1996). Auld and

Steiner describe Smith as 'still after a century the undisputed doyen of biblical geography' (p. 2).

107. J. Denney, 'Jerusalem', *The British Weekly*, 28 May 1908, p. 177.

108. Review of *Jerusalem* in the *Times Literary Supplement*, 28 May 1908 (unattributed – collected in NLS Acc 9446 No 137).

109. W. Robertson Nicoll to GAS: 'I feel I must not lose a moment of time in congratulating you on your truly monumental and magnificent work. Do you not think the time has come for revising the HGHL and making such additions as you may think proper? What we wish is to issue it uniform with 'Jerusalem' in two volumes' (29 April 1908; NLS Acc 9446 No. 142).

110. HGHL, p. xviii.

111. *Jerusalem*, Vol. I, p. xiv.

112. Lockhead, p. 99

113. Butlin, 'George Adam Smith', p. 392

114. So Smith, in the preface to the first edition of the *Historical Atlas of the Holy Land*, 1916, p. vii. The delay was due to 'other literary works and the duties of my office'. In 1935 he produced a second edition along with John Bartholomew, the son of John G. Bartholomew.

115. Butlin, 'George Adam Smith', p. 391.

116. Gardiner, *Bartholomew.*, p. 52.

117. ibid., p. 53.

118. Cook, 'Sir George Adam Smith', p. 334.

119. A.B. Bruce 'The Rev George Adam Smith DD, LLD', *The British Weekly*, 30 July 1896, where Bruce cites several works which took a detailed look at subjects related to the geography of Palestine.

120. ibid.

121. O. Palmer Robertson, *Understanding the Land of the Bible: A Biblical-Theological Guide*, (New Jersey, 1996) p. 1.

122. Max Miller, 'Old Testament History and Archaeology', *Biblical Archaeologist*, 50/1 (March 1987), p. 56.

123. Y. Ben-Arieh, *Rediscovery of the Holy Land*, p. 226.

124. Keith W. Whitelam, *The Invention of Ancient Israel, the silencing of Palestinian history*, (London, 1996) p. 41.

125. ibid., p. 42.

126. Philip R. Davies, 'In Search of "Ancient Israel"', (Sheffield, 1992) *Journal for the Study of the Old Testament* Supplement Series No 148, p. 31. See

Kaiser, *History of Israel*, for a discussion of the issues raised here.
127. V. Philips Long, 'The Art of Biblical History' in M. Silva (ed), *Foundations of Contemporary Interpretation*, (Illinois, 1996) p. 370.
128. *The Book of the Twelve Prophets*, Vol. I, p. 313.


SELECT BIBLIOGRAPHY


A.G. Auld and M. Steiner, *Jerusalem I: From the Bronze Age to the Macabees (Cities of the Biblical World)*, Cambridge: Lutterworth Press, 1996.

A.R.H. Baker and D. Gregory (eds.), *Explorations in Historical Geography: Interpretative Essays*, Cambridge: Cambridge University Press, 1984.

J.R. Bartlett (ed.), *Archaeology and Biblical Interpretation* London: Routledge, 1997.

Y.Ben-Arieh, *The Rediscovery of the Holy Land in the Nineteenth Century* Jerusalem: Magnes Press, 1983.

J.S. Black and G. Chrystal, *Lectures and Essays of William Robertson Smith*, London: Hodder and Stoughton, 1912.

E.H. Brown (ed.) *Geography Yesterday and Tomorrow* Oxford: Oxford University Press, 1980.

R. Butlin, 'George Adam Smith and the historical geography of the Holy Land: contents, contexts and connections', *Journal of Historical Geography*, 14.4 (1988).

—, *Historical Geography: Through the Gates of Space and Time* (London, 1993).

A.C. Cheyne, *Studies in Scottish Church History*, Edinburgh: T&T Clark, 1999.

P.R. Davies, *In Search of 'Ancient Israel'*, Sheffield: Sheffield University Press, 1992.

G.W. Grogan, 'Heilsgeographie: Geography as a Theological Concept', *The Scottish Bulletin of Evangelical Theology*, Volume 6, Number 2, Autumn 1988.

H.F. Hahn, *The Old Testament in Modern Research*, London: SCM Press, 1956.

W. Johnstone (ed.), *William Robertson Smith: Essays in Reassessment*, Sheffield: Sheffield University Press, 1995.

W.C. Kaiser, *A History of Israel from the Bronze Age to the Jewish Wars*, New York: Broadman and Holman, 1998.

W.S. Lasor, D.A. Hubbard and F.W. Bush, *Old Testament Survey* Grand Rapids: Eerdmans, 1994.

G.F. Owen, *Abraham to Allenby*, Grand Rapids: Eerdmans, 1939.

E.N. Lochhead, 'Scotland as the Cradle of Modern Academic Geography in Britain', *Scottish Geographical Magazine*, 97 (1981).

O.P. Robertson, *Understanding the Land of the Bible: A Biblical-Theological Guide*, New Jersey: Presbyterian and Reformed, 1996.

J.M. Schwartz, 'The Geography Lesson: photographs and the construction of imaginative geographies', *Journal of Historical Geography*, 22.1 (1996.

N. Shepherd, *The Zealous Intruders: The Western Rediscovery of Palestine*, London: Collins, 1987.

M. Silva (ed), *Foundations of Contemporary Interpretation* Illinois: Apollos, 1996.

G.A. Smith, *The Book of Isaiah*, Vol I, London: Hodder and Stoughton, 1888.
—, *The Book of Isaiah*, Vol II, London: Hodder and Stoughton, 1892.
—, *The Historical Geography of the Holy Land*, London: Hodder and Stoughton, 1894.
—, *The Book of the Twelve Prophets commonly called the minor*, Vol I, London: Hodder and Stoughton, 1896.
—, *The Book of the Twelve Prophets commonly called the minor*, Vol II, London: Hodder and Stoughton, 1898.
—, *The Life of Henry Drummond*, London: Hodder and Stoughton, 1899.
—, *Jerusalem: The Topography, Economics and History from the earliest times to* AD *70*, Vol I, London: Hodder and Stoughton, 1907.
—, *Jerusalem: The Topography, Economics and History from the earliest times to* AD *70*, Vol II, London: Hodder and Stoughton, 1908.
---- *Atlas of the Historical Geography of the Holy Land*, London: Hodder and Stoughton, 1915.
—, *Syria and the Holy Land*, London: Hodder and Stoughton, 1918.
—, *The Book of Deuteronomy*, Cambridge: Cambridge University Press, 1918.
—, *Jeremiah*, London: Hodder and Stoughton, 1923.

L.A. Smith, *George Adam Smith: A Personal Memoir and Family Chronicle*, London: Hodder and Stoughton, 1943.

D. Sutherland, 'The Interface between Theology and Historical Geography', *The Scottish Bulletin of Evangelical Theology*, Vol. XI, Number 1, Summer 1993

A.J.P. Taylor, *The First World War*, London: Penguin, 1963.

C.M. Watson, *Palestine Exploration Fund: Fifty Years' Work in the Holy Land; a Record and Summary 1865-1915*, London: Committee of PEF, 1915.
K.W. Whitelam, *The Invention of Ancient Israel, the silencing of Palestinian history*, London: Routledge, 1996.

## CHAPTER TWELVE

# Politics and the Travels of Gertrude Bell (1868-1926)

## Richard Long

*Introduction*

Gertrude Bell was a remarkable and effortless traveller in the Middle East. She was also a superb writer and a self-taught art critic, archaeologist and architect. Chance, and British Intelligence, took her to Iraq, where she spent over nine almost continuous and largely unrewarding years. This paper discusses her travels, her writing, the charge that she was a spy even before the First World War, and her chequered career as a diplomat.

*The Traveller*

Before 1914, Gertrude Bell made (according to my count) 15 journeys into the Muslim world. She began with a visit to Istanbul in 1888 and an 1892 sojourn in Persia; in 1893 she stayed with a relative in Algiers. In 1897-8 she went round the world with her brother Maurice, calling in on Egypt and Turkey, revisiting the latter in 1899. At the end of that year and into 1900, basing herself on Jerusalem (where she took intensive lessons in both Arabic and Persian), she travelled by camel in Turkey, Palestine, Jordan, Syria and Lebanon. She was in Algiers again in 1902, in which year she also spent time in Turkey, Lebanon, Syria and Palestine (where she continued her Arabic and Persian studies), and started off round the world once more, spending much time in India. Her 1905 journey took her from Beirut to Jerusalem, Jordan and Syria. She stayed in Egypt in 1906, and in 1907 returned to Turkey, where she had her fateful meeting in Konya with Dick Doughty-Wylie, the British Vice-Consul there. It was from this time that, like Lady Hester Stanhope, 'whom she increasingly resembled', she injected a new element into her travels: consolation for her unfulfilled love of the gallant soldier diplomat. This factor – first evident in the 1909 journey which had archaeologising at Ukhaydhir in Iraq as its focal point and included as stopping-off places Egypt, Syria and Turkey – perhaps accounted for the superiority of *Amurath to Amurath* over her other books. In 1911, she was in Ukhaydhir, Babylon and Carchemish, where she had a less fateful meeting with T.E. Lawrence, who she thought would 'make a traveller'.[2] She revisited Persia in 1913, and in 1913-14 undertook her famous journey to Hayil which led unerringly to the climax and failure of her career as a diplomat in Baghdad.

*The Writer*

Gertrude Bell's first stay in Persia inspired her to produce a translation of Hafiz, which won high praise from Professor Arberry, and a piece of juvenilia, self-conscious and pretentious in style, entitled *Persian Pictures*. *The Desert and the Sown*, published thirteen years late, in 1907, a blow-by-blow account of her 1905 journey, is her second-best travelogue, but has not much in the way of unusual literary merit. It is with *Amurath to Amurath*, published in 1911, that she produced a work which deserves to live. Not only is it an account of a more ambitious journey. While recounting the minutiae of her progress, like its predecessor, it now attains an elevation of style and, in particular, a mastery of dialogue – with brilliant reproduction in English of spoken Arabic – which propel it into a higher class of writing and endow it with qualities for which, like her letters, it should be remembered.

In the Gertrude Bell archive there is a wealth of material which could have been, and deserves to be, put into book form. Indeed, but for the outbreak of the First World War, 'She would almost certainly, while it was fresh in her mind, have written the account of her journey to Ha'il...'[3] Instead – always an obsessive and over dutiful correspondent with her family – she became, briefly, a writer of love letters which have not been adequately praised and a composer of diplomatic despatches for which her style was not suited.

### The Amateur Spy?

Victor Winstone, the second of Bell's modern biographers, implies that she was ripe for recruitment by British Intelligence in 1896, at the age of 28: '...a woman', he says, 'with an ability to speak Arabic, the resourcefulness to survive long spells in the desert, and a keen understanding of archaeology and ancient architecture to justify her travels, was at that time[4] almost too good an intelligence prospect to be true.' A prospect she may indeed have been, though it would have taken a keen-eyed spy-recruiter to have noted her promise. Unfortunately for his assertion, although she had by 1896 taken several sets of Arabic lessons, she had not yet made a desert journey or embarked on her spell as an archaeologist in Turkey and Iraq, a country she did not visit until 1909, when she was 41. The latest commentator on her, Rosemary O'Brien, gives a date to her claimed actual recruitment: 'Sometime in 1900,' she affirms, 'Bell took on a new role... fusing polite travel with an explicit form of information gathering... It is likely,' she adds, 'that Bell owed her debut as an informal agent to her friendship with a circle of empire boosters close to Foreign Secretary Edward Grey. At any rate, official documents confirm that before 1915, when she formally joined British Intelligence in Cairo at the Arab Bureau, "she had been in the [unpaid] employ of the Intelligence Division of the Admiralty."'[5] There is of course a large gap between 'Sometime in 1900' and 'before 1915', and there is little or no evidence that Bell did anything much which could justify the use of the term 'intelligence-gathering' before Hayil. In any case, having made her claim, O'Brien proceeds rapidly to undermine it. 'Whether to call Bell an actual spy matters little,' she continues. 'From 1900 until the Great War, she monitored the political pulse of Arabic Turkey; in the tents of powerful Arab sheikhs and the divans of Ottoman officials she spent hours sipping coffee, asking questions, and honing her insights into the state of affairs... London,' she adds, 'was curious to know just how far Germans had penetrated Turkish provinces in Syria, Mesopotamia, and north Arabia. More particularly, were they attempting to poison Arab minds against the British? On her many trips to the East in the first decade of the twentieth century, Bell tried to find answers to these questions for the Foreign Office in London.'[6]

O'Brien, I consider, enters the sphere of romance when she claims that 'Bell began to find that her budding interest in archaeology provided a useful cover for her presence in southern Mesopotamia..., where German excavators were at work.'[7] In the first place, she never troubled to acquire 'cover' for her activities in Turkey, Syria, Palestine or Iraq and would have scorned to do so. Even though Hugh, her father, was accused in 1916 of being pro-German,[8] the

German innuendo is a slur. Bell's stepmother, Florence, wrote one-act plays in German[9] and her diplomat relation, Frank Lascelles, was British ambassador in Berlin from 1896. It is true that she had many close dealings with German archaeologists and Alpine guides, but was military information likely to be found in Ukhaydhir, in Babylon or on the Finsteraarhorn? Even though, in addressing Austrians and Germans, she seems to have mostly used French, when in her letters to her family she makes use of a foreign phrase it is often a German one, with which her relations were clearly entirely comfortable.

Many of her diary entries and references in her letters support the view that before Hayil at least she was not a spy. One of her oldest friends, and her greatest,[10] was Valentine Chirol, known as 'Domnul', the deputy head, later head, of the Foreign Department of *The Times*. She met him in 1897/8 in Bucharest, where Lascelles was then Minister. In 1909, she remarked to her stepmother in a letter from Baalbek, 'As soon as I have learnt what is to be learnt in Aleppo, I shall write Mr Chirol a long letter about Syrian politics and if he doesn't think it worth publishing he can hand it on to you, if you like to see it.'[11] Ten weeks later, after staying in Baghdad with Col. Ramsay, the British Consul-General, she reported to her family that 'The result is that I have written a very serious letter to *The Times* about irrigation and railroads. I don't know whether Domnul will see fit to print it. The fact is one ends by feeling so much bound up with this country that one thinks all the time of what can be done to help it.' Twelve weeks later, from Istanbul and referring to the Young Turk revolution, she told her step-mother, 'I'm wondering if I shall write another letter to *The Times* on the thorny question of the counter revolution in the provinces.'[12]

It is my contention that, despite a friendship with Grey's private secretary, Willy Tyrrell, whom she knew as early as

1899 and who became Foreign Office Permanent Under Secretary in 1921, before 1915 Bell was no kind of official spy. Though her family and she were in easy contact with almost the highest in the land, it was to *The Times*, not to Whitehall, that she reported before Hayil. It was natural for an intelligent observer like her – keenly interested in a part of the world of which little was known in detail, and naive with regard to the use to which her observations might be put by less scrupulous players of the game – to send home accounts of everything she saw. But if you are a spy, you do not send them to *The Times* or refer your findings to unpaid honorary attachés like George Lloyd 'with whom', she told her stepmother, 'I want to have a long talk about Turkey, or with someone.'[13]

The most that one can safely concede before the conclusion of her Hayil journey – which, significantly, she accounted pointless, likely to result in only 'a few names added to the map'[14] – is that in 1909 an intelligence channel perhaps began to open. On 27 June that year, in a letter from Eregli in Turkey to her stepmother, Bell said that, once back in London, 'I should like to see Willie [Tyrrell] and tell him what's up in this country.' Nothing seems to have been formalised, however, for 18 months later she informed her that 'The political news I have sent to Domnul in a long letter which he will publish if he thinks fit.'[15]

Winstone perhaps goes too far with speculation that part at least of her Hayil journey was synchronised with an intelligence mission in Sinai undertaken by Lawrence and Leonard Woolley, archaeologists who mothballed their Carchemish site for the purpose and were about to be recruited by the Arab Bureau, as well as with an epic trans-Arabia journey by Captain Shakespear, the British consul in Kuwait, who also stopped at Hayil. He supports it with a photograph Bell took of Huwaytat tribeswomen, one of whom – it is true – looks remarkably like Lawrence.[16]

## The Diplomat

Gertrude Bell's journey to Hayil – which she embarked on in order to take her mind off Doughty-Wylie, a married man who displayed no disposition to leave his wife, and which Elizabeth Monroe claims would, but for her house-arrest in the Al Rashid capital, have included the seat of Ibn Sa'ud's power, Riyadh[17] – won her the Royal Geographical Society's gold medal and led directly to her diplomatic career. It was when the First World War was in its second year that the call came to her to put her experience and knowledge of the eastern Arab world to official use. Following a letter which found its way to Grey and differed from a possible pre-publication version of a successor to *Amurath to Amurath* in that it discussed the attitude to the Turks, Britain and France of the Arabs of Mesopotamia and Syria and the potential influence in the region of Ibn Sa'ud and the rulers of Kuwait and Muhammarah,[18] an invitation to join the Arab Bureau in Cairo came from D.G. Hogarth, who planned the Bureau and became its Executive Director.

Though she appears to have got on well enough with her all-male colleagues during her stay in Cairo, Hogarth later claimed that the work she did ensured 'much of the success of the Arab Revolt' through her provision of tribal information which Lawrence 'made signal use of in the Arab campaigns of 1917 and 1918',[19] they did not let much time pass before agreeing to a suggestion from the Viceroy of India, Lord Hardinge, a friend of Gertrude's also first met in Bucharest,[20] that she visit Delhi. Her mission – an unlikely one for so inexperienced an official – was to attempt to effect some co-ordination between the Viceroy and the British Consul-General in Cairo, Sir Henry McMahon, over the policies Britain was preparing to adopt in those *vilayetler* of the Ottoman Empire where it was anticipated that the Arabs would be free of Turkish shackles at the end of the First World War. She believed that in her three and a half weeks in India she had gone some way to fulfilling the task she had been set. She had not, however, been in Delhi very long when proposals began to be made to her that she should not rejoin the Arab Bureau in Egypt but become its temporary correspondent in Mesopotamia, which an Indo-British force had invaded in November 1914. Accordingly, it was in the unkind climate of Basrah that she disembarked, to find herself again the only woman among a group of British officials who were employed by the Foreign Department of the Government of India and were waiting to take up diplomatic positions in what was to become the capital of the new country of Iraq.

Bell joined Indian Expeditionary Force 'D' as the Arab Bureau's correspondent to Cairo[21] with Assistant Political Officer rank.[22] Almost a year after General Townshend and his men surrendered at Kut al 'Amarah, and a month after Sir Stanley Maude took Baghdad, she transferred to the capital. There she enjoyed three years of partial diplomatic success, followed by six of disappointment. Kathryn Tidrick correctly claims[23] that she was quite unsuited to be a proconsul[24] and Elie Kedourie was undoubtedly right in his comment that she proved herself far too emotional for the role, not only in relation to her British masters and, crucially, to Arnold Wilson, the second of her chiefs, but also towards the Iraqis themselves. She told her father in 1922, 'I'm happy in feeling that I've got the love and confidence of a whole nation.'[25]

Things went fairly well for her until Sir Percy Cox, who as Civil Commissioner was second in the hierarchy after the General Officer Commanding, left Baghdad in February 1918 to be Minister in Tehran for 32 months. With his departure, Arnold Wilson, known as A.T., acted as Civil Commissioner. Out of contempt for Lawrence in particular, he had no good opinion of the Arab Bureau, which was set up in order to influence British policy towards the liberated Arabs. From Wilson's standpoint, 'it sought to impose King Husain and his family upon the whole of Arabia, and at no time did its directors show any desire to look at the problems of Iraq from any other angle.'[26] The extremely high opinion Bell formed of him at the start[27] was not to last. Her attendance at Churchill's Cairo Conference brought about a 'conversion... to a belief in the desirability of a Sherifian solution in Iraq'[28] which – voiced in a report with the oddly sentimental name for a diplomatic document of *Syria in October*,[29] – was anathema to Wilson.[30] Their dispute was deepened by Bell's private correspondence with leading figures at home.

The combination of Bell's compulsive letter-writing and the high level connections of her family constituted a time bomb which Wilson was unable to defuse. She had had an arrangement about it with Cox – one which he seems to have implemented with some leniency[31] – which she made no attempt to adhere to under Wilson. In August 1918, he discovered[32] that she was writing behind his back.[33] In November, she suggested to Edwin Montagu, the Secretary of State for India, that Wilson should drop down from second to third in the hierarchy.[34] As Cox's return came closer, her irresponsibility grew. Early in 1920, she told her father that she had written privately[35] to Montagu – 'an immense letter about the sort of government we ought to set up here and even sent him the rough draft of a constitution'[36] – and a similar one to Sir Arthur Hirtzel, Wilson's immediate India Office superior.[37] She informed her stepmother that 'I'm in a minority of one in the Mesopotamian political service – or nearly – and yet I'm so sure I'm right that I would go to the stake for it... But they must see, they must know at home. They can't be so blind as not to read such gigantic writing on the wall...'[38] In the end, Wilson 'specifically demanded her recall... on the ground of her mischief-making...'[39] and Cox, whom she had earlier protected against Maude, had to come to her aid.⁴⁰

Bell's lack of self-discipline and her petulant behaviour left Wilson in the unenviable position of having a member of his staff who saw no need to hide from influential outsiders not only the fact that she did not agree with him but also her belief that she did not consider that he should be heading the British mission. With a complete lack of logic, she resented Wilson's occupancy of the post and regarded him as having usurped Cox's position, 'he being my real Chief'.⁴¹

Making things uncomfortable for Wilson and herself was one thing. Allowing her differences with her boss to affect policy on the ground was quite another. In June 1920 she admitted to her father, 'most unfortunately I gave one of our Arab friends here a bit of information I ought not technically to have given.' Wilson's reaction was that 'my indiscretions were intolerable, and that I should never see another paper in the office… "You've done more harm than anyone here",' he said. She commented, 'I know really what's at the bottom of it – I've been right and he has been wrong.'⁴²

Closely following on one in Egypt, a rebellion broke out in Iraq in July 1920. On the first of the month, Gertrude inaugurated her politically uninfluential and embarrassing final period in Baghdad by writing privately to the G.O.C., Gen. Sir Aylmer Haldane, without informing Wilson. She told him, 'The bottom seems to have dropped out of the agitation and most of the leaders seem only too anxious to let bygones be bygones.' In his account of this, Wilson said that Haldane used her view to justify 'the unduly optimistic attitude he took during the early stages of the rebellion',⁴³ which led him, when deploying some of his troops, to neglect at least one important trouble-spot. As a consequence, he wrote on 14 July to Cox, now in London, to try for a second time to get rid of her.⁴⁴ Her indiscretions harmed her even in the eyes of her adored first chief. Misleading the G.O.C. ensured that, on his return to Baghdad, Cox and – on his final departure – his successor

Sir Henry Dobbs would deny Bell a place at the political top table. She had played a major part in drawing up the mandate for Iraq which came into force in April, 1920. Now she had to look on as Cox, King Faysal and Hubert Young drew up the 1922 British-Iraqi treaty. With justice, she confessed to her parents, 'I think I may have been of some use here but I suspect I've come very near the end of it.'[45] She was not only a spent force in relation to her colleagues. The Iraqis, too, had become disillusioned with her, even Nuri as Sa'id and King Faysal, both of whom she had more than admired, in the end finding her a nuisance.[46]

*Conclusion*

Gertrude Bell's tragedy was that her incessant travelling and the accounts she wrote of it combined with the onset of war to divert her talents into avenues where they could no longer flourish. Her nine years of largely inappropriate effort in Iraq represented the loss of someone who might have become an even greater traveller and an even finer writer than she had been before the Arab Bureau beckoned her down the wrong track.

REFERENCES

1. Tidrick, Kathryn, *Heart Beguiling Araby, the English Romance with Arabia*, I.B. Tauris, London, 1989, p. 188.
2. Bell, Lady (ed.), *The Letters of Gertrude Bell*, Ernest Benn, London, 1927, pp. 305-6.
3. Burgoyne, Elizabeth, *Gertrude Bell from her Personal Papers, 1914-1926*, Ernest Benn, London, 1961, p. 13, supported by two 26 March 1914 entries in the diary Gertrude wrote for Doughty-Wylie, in Burgoyne, Elizabeth, *Gertrude Bell from her Personal Papers, 1889-1914*, Ernest Benn, London,1958, p. 303.

4. When Professor Denison Ross was head of the London University School of Oriental Studies and special adviser to the Directorate of Military Intelligence. (Winstone, H.V.F., *The Illicit Adventure*, Jonathan Cape, London, 1982, pp. 41-2).
5. O'Brien, Rosemary, Gertrude Bell, *The Arabian Diaries*, Syracuse University Press, New York, 2000, p. 11.
6. Bell, Lady, op. cit., p. 252.
7. O'Brien, op. cit., p. 12.
8. Burgoyne 1961, p. 45.
9. Winstone, H.V.F., *Gertrude Bell*, Constable, London, 1993, p. 11.
10. 11 December 1913 letter from Gertrude to Chirol, in Burgoyne 1958, p. 284.
11. 2 February, 1909.
12. 7 July, 1909.
13. 7 July, 1909.
14 16 February 1914 diary entry for Doughty-Wylie, in Burgoyne 1958, p.296.
15. 27 January 1911, from Damascus.
16. Winstone, H.V.F., *Woolley of Ur*, Secker and Warburg, London, 1990, p. 53.
17. Monroe, Elizabeth, *Philby of Arabia*, Ithaca Press, Reading, 1998, p. 45.
18. Details in Burgoyne 1961, pp. 14-15
19. Wallach, Janet, *Desert Queen*, Weidenfeld and Nicolson, London, 1996, p. 202.
20. Burgoyne 1958, p. 18.
21 Burgoyne 1961, p. 43.
22. Gertrude letter to Chirol, 13 September 1917, in Burgoyne 1961, p. 64.
23. Tidrick, op. cit., p. 189.
24. 'such duties require a toughness of mind and spirit she did not possess.'
25. From Baghdad, 16 February.
26. Wilson, Sir Arnold Talbot, *Mesopotamia, 1917-1920, a Clash of Loyalties*, O.U.P., 1931, p. 110. 'From our point of view in 'Iraq', he said, 'the existence of the Arab Bureau was an embarrassment rather than an advantage: the relative proximity of Cairo to Paris and London gave the exponents of the Hashimite policy, themselves for the most part untrammelled by office or by administrative responsibilities, an advantage

over the accredited representatives of the British Government in 'Iraq and in the Persian Gulf.'
27. For this, see Burgoyne 1961, pp. 84-5, Marlowe, John, *Late Victorian, The Life of Sir Arnold Talbot Wilson*, The Cresset Press, London, 1967, p.129, and Gertrude's letter from Baghdad to her parents, 24 May 1918.
28. Marlowe, op. cit., p. 202.
29. Marlowe, op. cit., pp. 160-1.
30. Winstone 1993, pp. 217-8.
31. See the letter of 22 February 1918 he took to Lord Hardinge, in Burgoyne 1961, p. 78.
32. A Private MS communication from Sir Arthur Hirtzel, Political Secretary of the India Office, to Wilson, referring to 'a letter from Miss Bell which contains a flaming testimonial to your qualities and to the success of your administration...' (Wilson, Sir Arnold Talbot, Papers, British Library: Add Mss 52455C).
33. Winstone 1993, p. 209.
34. Marlowe, op. cit., p. 202.
35. 4 January letter, in Burgoyne 1961, p.124, Marlowe, op. cit., p. 202.
36. Wallach, op. cit., p. 248.
37. Marlowe, op. cit., p. 202.
38. 12 January 1920, from Baghdad.
39. Philby, H. St. J.B., *Arabian Days*, Robert Hale, London, 1948, p. 185.
40. Marlowe, op. cit., p. 203, Wilson to Gertrude, in Wallach, op. cit., p. 261.
41. Gertrude letter from Baghdad to her father, 20 June 1920.
42. Gertrude letter 14 June 1920.
43. Wilson, Sir Arnold Talbot, *Loyalties, Mesopotamia 1914-1917*, OUP, 1930, p. 275n.
44. 'If you can find a job for Miss Bell at home, I think you would be well-advised to do so. Her irresponsibilities are a source of considerable concern to me and are not a little resented by Political Officers' (Marlowe, op. cit., p.205).
45. 1 November 1921, from Baghdad.
46. De Gaury, Gerald, *Three Kings in Baghdad*, Hutchinson, London, 1961, p. 44.

# BIBLIOGRAPHY

Bell Archive, www.gerty.ncl.ac.uk

Bell, Gertrude, *Amurath to Amurath*, Macmillan, London, 1911.
—, *The Desert and the Sown*, Heinemann, London, 1907.
—, *Persian Pictures*, Ernest Benn, London, 1894.
—, Bell, Lady (ed.), *The Letters of Gertrude Bell*, Ernest Benn Limited, 1927.

Brittain, Vera, *The Women at Oxford*, Harrap, London, 1960.

Burgoyne, Elizabeth, *Gertrude Bell from her Personal Papers*,
(1) 1889-1914, Benn, London, 1958
(2) 1914-1926, Benn, London, 1961.

De Gaury, Gerald, *Three Kings in Baghdad*, Hutchinson, London, 1961.

Gordon, Lesley, *Gertrude Bell*, Newcastle upon Tyne, 1994.

Graves, Philip, *The Life of Sir Percy Cox*, Hutchinson, London, 1941.

Ireland, Philip W., *'Iraq, a Study in Political Development*, Cape, London, 1937.

Kedourie, Elie, *The Chatham House Version and other Middle Eastern Studies*, University Press of New England, London, 1984.

Marlowe, John, *Late Victorian, The Life of Sir Arnold Talbot Wilson*, The Cresset Press, London, 1967.
Monroe, Elizabeth, *Philby of Arabia*, Ithaca Press, Reading, 1998.

O'Brien, Rosemary, *Gertrude Bell, the Arabian Diaries*, Syracuse University Press, N.Y., 2000.

Philby, H. St. J.B., *Arabian Days*, Robert Hale, London, 1948.

Tidrick, Kathryn, *Heart Beguiling Araby, the English Romance with Arabia*, I.B. Tauris, London, 1989.

Wallach, Janet, *Desert Queen*, Weidenfeld and Nicolson, London, 1996.

Wilson, Sir Arnold T., *Papers*, British Library: Add Mss 52455C
—, *Loyalties, Mesopotamia 1914-1917*, OUP, 1930.
—, *Mesopotamia, 1917-1920, a Clash of Loyalties*, OUP, 1931.

Wingate, Sir Ronald, *Not in the Limelight*, Hutchinson, London, 1959.

Winstone, H.V.F., *Gertrude Bell*, Constable, London, 1993.
—, *The Illicit Adventure*, Jonathan Cape, London, 1982.
—, *Woolley of Ur*, Secker and Warburg, London, 1990.

Printed in the United Kingdom
by Lightning Source UK Ltd.
118417UK00001B/160-195